# PHYSICAL AND CHEMICAL METHODS IN SOIL ANALYSIS

## Fundamental Concepts of Analytical Chemistry and Instrumental Techniques

(SECOND EDITION)

# PHYSICAL AND CHEMICAL METHODS IN SOIL ANALYSIS

## Fundamental Concepts of Analytical Chemistry and Instrumental Techniques

**(SECOND EDITION)**

**DIPAK SARKAR**
MSc, PhD, FIC, FASCT, FISSS, FNESA
Former Director
National Bureau of Soil Survey and
Land Use Planning (ICAR)
Amravati Road, Nagpur

**ABHIJIT HALDAR**
MSc, PhD
National Bureau of Soil Survey and
Land Use Planning (ICAR)
Regional Centre
Kolkata

NEW AGE
REFERENCE

*An Imprint of*

**NEW AGE INTERNATIONAL (P) LIMITED, PUBLISHERS**
**LONDON • NEW DELHI • NAIROBI**
Bangalore • Chennai • Cochin • Guwahati • Hyderabad • Kolkata • Lucknow • Mumbai
Visit us at **www.newagepublishers.com**

Published by New Age International (P) Ltd., Publishers
First Edition: 2005
Second Edition: 2010
**Reprint: 2016**



## GLOBAL OFFICES

- **New Delhi** **NEW AGE INTERNATIONAL (P) LIMITED, PUBLISHERS**
  7/30 A, Daryaganj, New Delhi-110002, (INDIA)
  **Tel.:** (011) 23253771, 23253472, **Telefax:** 23267437, 43551305
  **E-mail: contactus@newagepublishers.com • Visit us at www.newagepublishers.com**
- **London** **NEW AGE INTERNATIONAL (UK) LTD.**
  27 Old Gloucester Street, London, WC1N 3AX, UK
  **E-mail: info@newacademicscience.co.uk • Visit us at www.newacademicscience.co.uk**
- **Nairobi** **NEW AGE GOLDEN (EAST AFRICA) LTD.**
  Ground Floor, Westlands Arcade, Chiromo Road (Next to Naivas Supermarket)
  Westlands, Nairobi, KENYA, **Tel.:** 00-254-713848772, 00-254-725700286
  **E-mail: kenya@newagepublishers.com**

## BRANCHES

- **Bangalore** 37/10, 8th Cross (Near Hanuman Temple), Azad Nagar, Chamarajpet, Bangalore- 560 018
  **Tel.:** (080) 26756823, **Telefax:** 26756820, **E-mail: bangalore@newagepublishers.com**
- **Chennai** 26, Damodaran Street, T. Nagar, Chennai-600 017, **Tel.:** (044) 24353401, **Telefax:** 24351463
  **E-mail: chennai@newagepublishers.com**
- **Cochin** CC-39/1016, Carrier Station Road, Ernakulam South, Cochin-682 016, **Tel.:** (0484) 2377303, **Telefax:** 4051304
  **E-mail: cochin@newagepublishers.com**
- **Guwahati** Hemsen Complex, Mohd. Shah Road, Paltan Bazar, Near Starline Hotel, Guwahati-781 008,
  **Tel.:** (0361) 2513881, **Telefax:** 2543669, **E-mail: guwahati@newagepublishers.com**
- **Hyderabad** 105, 1st Floor, Madhiray Kaveri Tower, 3-2-19, Azam Jahi Road, Near Kumar Theater, Nimboliadda
  Kachiguda, Hyderabad-500 027, **Tel.:** (040) 24652456, **Telefax:** 24652457
  **E-mail: hyderabad@newagepublishers.com**
- **Kolkata** RDB Chambers (Formerly Lotus Cinema) 106A, 1st Floor, S N Banerjee Road, Kolkata-700 014
  **Tel.:** (033) 22273773, **Telefax:** 22275247, **E-mail: kolkata@newagepublishers.com**
- **Lucknow** 16-A, Jopling Road, Lucknow-226 001, **Tel.:** (0522) 2209578, 4045297, **Telefax:** 2204098
  **E-mail: lucknow@newagepublishers.com**
- **Mumbai** 142C, Victor House, Ground Floor, N.M. Joshi Marg, Lower Parel, Mumbai-400 013
  **Tel.:** (022) 24927869, **Telefax:** 24915415, **E-mail: mumbai@newagepublishers.com**
- **New Delhi** 22, Golden House, Daryaganj, New Delhi-110 002, **Tel.:** (011) 23262368, 23262370, **Telefax:** 43551305
  **E-mail: sales@newagepublishers.com**

**ISBN: 978-81-224-2725-7**
**C-16-07-9537**

Printed in India at Glorious Printers, Delhi.
Typeset at In-house, Delhi.

**NEW AGE INTERNATIONAL (P) LIMITED, PUBLISHERS**
7/30 A, Daryaganj, New Delhi-110002
Visit us at **www.newagepublishers.com**
**(CIN: U74899DL1966PTC004618)**

# Foreword

Agriculture is the mainstay of the economy of our country and only the sustainable agriculture is likely to provide long term food production, development and poverty alleviation. Modern civilization is facing a real threat from the rapid population outburst. Simultaneously the *per capita* land area or land-man ratio is decreasing dangerously which is one of the main reason for food insecurity in the near future. Since soil is the backbone of any civilization and is the most precious and vital natural resource, it must be thoroughly understood and conserved/managed well for sustained agricultural production.

The present textbook is a comprehensive analytical manual covering the aspects of soil analysis in the major areas of Soil Physics and Soil Chemistry. Furthermore, the concept of soil microbial biomass carbon and nitrogen is also dealt in detail. An important feature of this text is that it describes not only the analytical procedures in detail but also furnishes sufficient theoretical background on the subject matter. The fundamental principles of the analytical methods have been discussed precisely and the theories are explained well with mathematical analysis and chemical reactions wherever required.

I hope that this textbook will be of very much use for the undergraduate and post graduate students of Agricultural Universities/Institutes in India, researchers, teachers and those interested in the analytical study of the soil.

Finally I appreciate the authors' untiring effort in giving shape to this present text.

I wish them all success in their endeavour.

**Former Professor & Head** *S.K. Gupta*
*Division of Agricultural Chemistry and*
*Soil Science, University of Calcutta*
*35, Ballygunge Circular Road*
*Kolkata–700 019*

**Former President**
*Agricultural Sciences Section,*
*Indian Science Congress Association, 2000*

# Preface to the Second Edition

This text is primarily meant to cater the need of undergraduate and postgraduate students of Agricultural Universities/Institutes in India and is expected to be of help to teachers and researchers as well. An endeavour has been made to provide sufficient theoretical background on the subject matter to ensure that the procedures are not followed merely to obtain a numerical answer.

The text comprises of four major areas viz. Soil Physics, Soil Chemistry, Fundamental Concepts of Instrumental Techniques and Fundamental Concepts of Analytical Chemistry. Each topic is presented in a lucid and concise manner furnishing details of reagent preparation and stepwise procedure, outlining precautions and additional notes wherever necessary. The principles have been discussed briefly and theories explained well with mathematical derivations and chemical equations as and when required. The analytical methods described in this text are either being widely used or have been accepted throughout as standard. Various methods have been explained in a simple and easily understandable language comprising of principle with equipments and apparatus, procedure, observations and calculations.

The second edition is improved with some important additions in the section of Soil Chemistry particularly in relation to environment. Ground water and soil contamination with arsenic, selenium, fluoride and other heavy metals viz. lead, cadmium, chromium, nickel, has reached alarming levels in many parts of our country endangering human health and environment. A comprehensive account of estimation of such soil pollutants is dealt in details along with required theoretical back up in the subject.

Inspite of best efforts by the authors, the text may still have some discrepancies. Suggestions for improvement from the readers will be highly appreciated.

*National Bureau of Soil Survey*
*& Land Use Planning (ICAR)*
*Amravati Road,*
*Nagpur – 440 010*

**Dipak Sarkar**
**Abhijit Haldar**

# Preface to the First Edition

This text is primarily meant to cater the need of undergraduate and postgraduate students of Agricultural Universities/Institutes in India and is expected to be of help to teachers and researchers as well. An endeavour has been made to provide sufficient theoretical background on the subject matter to ensure that the procedures are not followed merely to obtain a numerical answer.

The text comprises of four major areas viz. Soil Physics, Soil Chemistry, Fundamental Concepts of Instrumental Techniques and Fundamental Concepts of Analytical Chemistry. Each topic is presented in a lucid and concise manner furnishing details of reagent preparation and stepwise procedure, outlining precautions and additional notes wherever necessary. The principles have been discussed briefly and theories explained well with mathematical derivations and chemical equations as and when required. The analytical methods described in this text are either being widely used or have been accepted throughout as standard. Various methods have been explained in a simple and easily understandable language comprising of principle with equipments and apparatus, procedure, observations and calculations.

In spite of best efforts by us, the text may still have some discrepancies. Suggestions for improvement from the readers will be highly appreciated.

*National Bureau of Soil Survey*
*& Land Use Planning (ICAR)*
*Amravati Road,*
*Nagpur – 440 010*

**Dipak Sarkar**
**Abhijit Haldar**

# Acknowledgements

The authors express their deep sense of gratitude to the following persons for their encouragement, help, co-operation and assistance in various capacities at different stages during bringing out this document.

- Dr. K.S. Gajbhiye, Director, National Bureau of Soil Survey and Land Use Planning (Indian Council of Agricultural Research), Nagpur for encouragement and support.
- Dr. Utpal Baruah, Principal Scientist, National Bureau of Soil Survey and Land Use Planning (Indian Council of Agricultural Research), NER Centre, Jorhat for constant support.
- Professor Shyamal Kumar Gupta (Retd.), University of Calcutta and Professor Saroj Kumar Sanyal, Bidhan Chandra Krishi Viswavidyalaya, Mohanpur, Nadia, West Bengal for their inspiration and support.
- The Scientists of Regional Centre, National Bureau of Soil Survey and Land Use Planning (Indian Council of Agricultural Research), Regional Centre, Kolkata specially Dr. D.S. Singh, Dr. A.K. Sahoo, Dr. K.D. Sah, Dr. K. Das, Dr. T.H. Das, Dr. D.C. Nayak, Dr. D. Dutta, Dr. S.K. Gangopadhyay, Shri S. Mukhopadhyay, Smt. T. Banerjee, Dr. T. Chattopadhyay for their constant support and encouragement with valuable suggestions time to time.
- Shri B.K. Saha, Smt. Nirmala Kumar, Shri B.C. Naskar, Shri Pranabesh Mondal, Shri Sourav Ghosh (Ex-SRF) and all others of National Bureau of Soil Survey and Land Use Planning (Indian Council of Agricultural Research), Regional Centre, Kolkata who rendered support and discharged their duties to accomplish the job.
- To all others who rendered their support to give the final shape to the document.

**Authors**

# Contents

# 1 Instrumental Techniques: Fundamental Concepts

## 1.1 pH: GENERAL DISCUSSION

pH was originally defined as $-\log\,(m_{H^+}/m^-)$ where $m_{H^+}$ = molality of $H^+$ and $m^-$ is unity *i.e.*, 1 mole $kg^{-1}$ (exactly) but later was defined in terms of activity (introduction of $m^-$ keeps the terms inside logarithm dimensionless).

**Sorensen** (1909) defined pH of a solution as the negative logarithm of the hydrogen ion activity, which in very dilute solutions can be expressed as concentration in g mole per litre.

$$pH = -\log_{10} a_{H^+} \text{ or } a_{H^+} = 10^{-pH} \qquad ...(1.1.1)$$

$a_{H^+}$ represents the activity of hydrogen ions — refers strictly to a true solution in which the ions are completely dissociated where there exists a large volume compared to molecular dimensions.

When solution is very dilute

$$pH = -\log_{10} C_{H^+} \qquad [\text{since } a_{H^+} = C_{H^+}] \qquad ...(1.1.2)$$

$$C_{H^+} = 10^{-pH}$$

Now, for a solution of pH = 4, $C_{H^+} = 10^{-4}$ and for a solution of pH = 9, $C_{H^+} = 10^{-9}$.

When concentrations are not low enough for molalities to be used, activity coefficients can be estimated from the Debye-Hückel, limiting law or its extended form which read as

$$\log_{10} r\pm = \frac{-Az^+z^-\sqrt{I}}{1 + aA_1\sqrt{I}} + A_2I \qquad ...(1.1.3)$$

where $z^+$ and $z^-$ are the numerical values of the valence of the two ions of the electrolyte, 'I', is the ionic strength, $a$ is the effective radius of ion particles or more appropriately closest distance between the ions; A and $A_1$ are constants given as

$$A = B/2.303 = 0.509 \text{ at } 25°C$$

where $$B = 1/(DT)^{3/2}\ (\varepsilon^2 N/R\ \sqrt{2\pi\varepsilon^2 ND/k\cdot 1000})$$

$$A_1 = \sqrt{\frac{(4\pi\,\varepsilon^2\,2Nd)}{DkT\cdot 100}}$$

$A_2$ is another constant which accounts for variation of dielectric constant or a constant for a given electrolyte.

$\varepsilon$ = electronic charge = $4.77 \times 10^{-10}$ e.s.u.

N = Avogadro's number = $6.023 \times 10^{23}$

$k$ = R/N = $8.314 \times 10^7/6.023 \times 10^{23}$

= $1.38 \times 10^{-16}$ ergs at 25°C = Boltzmann's constant

D = dielectric constant = 78.54

T = absolute temperature

$d$ = density of the solution, the same as that of solvent when the solution is dilute

The activity coefficients of equilibrium solution in soil chemistry studies are often determined using Davies equation (Amacher, 1984) namely,

$$\log r = \frac{(-0.502z^2\sqrt{I})}{(1+\sqrt{I})} = 0.2I \qquad ...(1.1.4)$$

where $z$ is the valency of an ion and I is ionic strength of the soil solution. The ionic strength is calculated from the electrical conductivity (ECe) according to the relation proposed by Griffin and Jurinak (1973), namely

$$I = 0.0127\ \text{ECe} \qquad ...(1.1.5)$$

***Note:*** In 1.0 mole kg$^{-1}$ HCl (aq), $m_{H^+}$ = 1.0 mole kg$^{-1}$ (the acid is fully ionised) and $\gamma\pm = 0.811$ (Table value, Atkins, 1986); therefore, $a_{H^+} = 0.811\times(1.0 \text{ mole kg}^{-1}/m^-) = 0.81$, implying pH = 0.092 in place of the value pH = 0 which would have been obtained from the use of molality alone. There is also nothing mysterious about the concept of negative pH, for it, merely corresponds to an activity greater than unity. For example, in 2.00 mole kg$^{-1}$ HCl (aq) where the mean activity coefficient is 1.011 (Table value, Atkins, 1986), the hydrogen ion activity is 2.02, implying pH = –0.31.

In the pure state, water is dissociated to a very small extent and behaves as a weak electrolyte. The equilibrium constant of the dissociation, $H_2O \rightleftharpoons H^+ + OH^-$, is given by

$$K = \frac{a_{H^+}\,.\,a_{OH^-}}{a_{H_2O}} \qquad ...(1.1.6)$$

In the pure state, or in dilute solution, the activity of water $a_{H_2O}$ is constant and is taken to be unity.

Hence,

$$K_w = a_{H^+}\,.\,a_{OH^-} \qquad ...(1.1.7)$$

The $K_w$ is called the ionic activity product of water. Replacing activities with concentrations and activity coefficients

$$K_w = C_{H^+} f_{H^+} \cdot C_{H^-} f_{H^-} = (C_{H^+} C_{H^-}) f_{H^+} \cdot f_{H^+} \qquad ...(1.1.8)$$

or

$$K_w = K'_w f_{H^+} \cdot f_{OH^-} \qquad ...(1.1.9)$$

where $K'_w = C_{H^+} \cdot C_{OH^-}$ is called the ion product of water. In pure water or in dilute solutions the activity coefficients $f_{H^+}$ and $f_{OH^-}$ are almost unity and so $K_w \approx K'_w$. That is no appreciable error is involved in accepting ion product of water as its ionic activity product.

At 25°C, the concentration of $H^+$ ions in pure water has been found to be $1 \times 10^{-7}$. Since $C_{H^+} = C_{OH^-}$ in pure water.

$$\therefore \qquad K'_w = C_{H^+} \cdot C_{OH^-} = (1 \times 10^{-7})^2 = 1 \times 10^{-14} \qquad ...(1.1.10)$$

The ionic activity product of water is very accurately derived, from e.m.f. measurement of suitable galvanic cells, such as

$$\underset{}{Pt\ (H_2)} \mid \underset{(m_1)}{KOH\ (aq)} \mid \underset{(m_2)}{KCl\ (aq)} \mid AgCl\ (s) \mid Ag;$$ ($m_1$ and $m_2$ are the molalities)

in which the cell reaction is, $AgCl\ (s) + ½H_2 \rightarrow Ag\ (s) + H^+ + Cl^-$. The experimentally obtained value from e.m.f. determination of $K_w$ was found to be $1.008 \times 10^{-14}$ at 25°C. The ionic activity product of water at different temperatures are:

| Temp (°C) | $K_w \times 10^{-14}$ |
|---|---|
| 0 | 0.114 |
| 18 | 0.578 |
| 25 | 1.008 |
| 40 | 2.919 |
| 50 | 5.344 |

It becomes evident from equation that $K_w$ or $K'_w$ is a temperature dependant quantity. Accordingly the $C_{H^+}$ and $C_{OH^-}$ will also vary with temperature thus making pH determination a temperature sensitive measurement.

Equation (1.1.9) really suggests that in an aqueous medium, the product of the concentrations of $H^+$ and $OH^-$ should be constant. If we are dealing not with pure water, but a dilute aqueous solution, this relation is still valid. In an acid solution, there is a preponderance of $H^+$ ions but nevertheless there would be some $OH^-$ ions and the product of two concentrations would be $1 \times 10^{-14}$ at 25°C. Similarly, in alkali solutions, there exists some $H^+$ ions. For instance, in (M/100) HCl solution

$$C_{OH^-} = \frac{K'_w}{C_{H^+}} = \frac{1 \times 10^{-14}}{1 \times 10^{-2}} = 1 \times 10^{-12} \qquad ...(1.1.11)$$

The value of ion product of water can be obtained experimentally from conductivity measurement of pure water and also from electromotive force measurement of some suitable galvanic cells. The value of $K_w$ was observed to be $1.008 \times 10^{-14}$ at 25°C from e.m.f. measurement. The value of $K_w$ is sometimes expressed in its logarithmic form, such that

$$pK_w = -\log K_w \quad ...(1.1.12)$$

At 25°C

$$pK_w = -\log (1 \times 10^{-14}) = 14 \quad ...(1.1.13)$$

Just as the way the pH has been defined, similarly, the activity of $OH^-$ ions is expressed in pOH scale defined as

$$pOH = -\log_{10} a_{OH^-} \quad ...(1.1.14)$$

or

$$a_{OH^-} = 10^{-pOH}$$

From equation (1.1.7)

$$a_{H^+} \cdot a_{OH^-} = K_w$$

or

$$(-\log a_{H^+}) \cdot (\log a_{OH^-}) = -\log K_w$$

or

$$pH + pOH = pK_w = \text{constant} \quad ...(1.1.15)$$

That is as pH increases, pOH must decrease and vice versa.

In pure water, which is neutral, $C_{H^+} = C_{OH^-} = 10^{-7}$ *i.e.*, pH of water is 7. Hence, the neutral solution has pH = 7. Any solution having pH lower than 7 will be acid and a solution having pH above 7 will be alkaline. Thus at 25°C, pH of 0.001 M KOH will be 9 for $C_{H^+} = K_w/C_{OH^-} = 10^{-14}/10^{-6} = 10^{-9}$ *i.e.*, pH = 9.

It is very cumbersome to express the concentrations of $H^+$ or $OH^-$ ions since the numerical values are extremely small; smallest being $10^{-14}$ which is $1/10^{14}$ moles per litre or 0.000,000,000,000,01. Sorensen therefore suggested the use of the negative logarithm values so that simple whole numbers are used. For example, if $C_{H^+} = 10^{-7}$ then log of $10^{-7} = -7 \times 1$ (since log 10 = 1). The negative of this value is 7. Thus the pH can be expressed in numerical values ranging from 0 to 14 as fixed points. The values below 7 indicates acidity and those above 7 indicates alkalinity.

***Note:*** Since the extent or degree of dissociation is temperature dependent, the pH scale (0–14) is valid for a particular temperature. For other temperature necessary adjustements are to be made.

### 1.1.1 Measurement of pH

The most accurate method of ascertaining the pH of a solution depends on e.m.f. (electromotive force) measurement. The given solution is made the electrolyte of a half cell such that its potential is governed by the $H^+$ ion concentration of the solution. This half cell is then coupled with a reference electrode and the e.m.f. of the cell measured potentiometrically. The different types of half cells or single electrodes commonly used are hydrogen electrode, quinhydrone electrode, glass electrode, antimony electrode, etc. In the conventional instruments the measuring electrode is of glass and the reference one is calomel electrode.

### 1.1.2 Glass Electrode

If a thin glass membrane separates two solutions a potential is developed, across the membrane. The magnitude of this membrane potential depends mainly on the pH of the solutions. If pH of one of the solution is kept constant and the other varied, then the electrode potential follows the relation:

$$\xi_G = \xi^{\circ}_G - RT/F \ln a_{H^+} = \xi^{\circ}_G + 2.303\ RT/F\ pH \qquad ...(1.1.2.1)$$

The glass electrode consists of a thin membrane of a specific prepared soft glass globe containing a dilute solution of hydrochloric acid in which is immersed Ag-AgCl electrode. The electrode is

Ag-AgCl (s) | 0.1 (N) HCl | Glass | Unknown solution ($a_{H^+}$)

The electrode potential of this half cell, is given in equation (1.1.2.1), in which $\xi_G$ includes a "small asymetry potential" which exists across the glass membrane due to internal strain. When this electrode is coupled with a reference electrode, say calomel electrode, the cell obtained is,

Ag-AgCl (s) | 0.1 (N) HCl | Glass | Unknown solution ($a_{H^+}$) | Standard calomel electrode

The e.m.f. of the cell is

$$E = \xi_G - \xi_{cal} = \xi^{\circ}_G + 2.303\ RT/F\ pH - \xi_{cal} \qquad ...(1.1.2.2)$$

In practice, the assembly of glass electrode is first used with a solution of known pH, say $pH_1$ and its e.m.f. is $E_1$. This solutions is then, substituted with the unknown solution so that

$$\Delta E = E_1 - E = 2.303\ RT/F\ (pH_1 - pH) \qquad ...(1.1.2.3)$$

It is thus immaterial what reference electrode is employed provided the same is used for both the measurements. The glass electrode and the reference electrode are suspended in the given solution and the e.m.f. of the cell measured with an electronic voltmeter. Ordinary potentiometer cannot be used due to the very high resistance of the glass membrane. A pH meter is actually a direct current amplifier that measures the e.m.f. which appears across the electrodes upon being immersed in a solution, soil suspension or irrigation water. The meter is graduated to read directly in pH units along with the e.m.f. (millivolts) scale. A standard, buffer solution of known pH is used to calibrate the instrument before determining the pH of the test solution. This is because an asymetric potential develops across the glass of the electrode even when it is immersed in a solution with a hydrogen ion concentration identical to that inside the bulb due to a difference in strain inside and outside the bulb.

### 1.1.3 Calomel Electrode

The electrode consists of mercury in contact with a solution of potassium chloride saturated with mercurous chloride. It maintains a constant potential, at a given temperature. In commercial models, a paste of mercury and $Hg_2Cl_2$ is contained in an inner tube connected to the KCl solution in an outer jacket. The lead wire is connected to the $Hg_2Cl_2$ paste through a column of mercury. The outer tube ends in a fine capillary to provide a salt bridge through the test solution to the glass electrode. pH meter with single (actually combined) electrode is also available as in case of digital type instrument.

The advantage of glass electrode is that it can be used in any solution not being affected by organic compounds or oxidising and reducing agents. A small quantity of solution is sufficient for determination of the pH. Special glass membranes are required when pH of the solution is very high (pH > 10). Such special electrodes are also commercially available where sodium of the glass is replaced by lithium. Most of the pH meter used in the soil testing laboratories in India, are Vacuum Type Volt Meters (VTVM). VTVM with indicating scales in pH values is calibrated in voltage units for a glass reference electrode pair on the basis of the relationship for the e.m.f. of pH cell. The apparent e.m.f./pH slope will be 59.15 mV per pH unit at 25°C using the equation pH = pHs-(E-Es)/0.000198T, where pHs and Es are the values in the standard state and T is the absolute temperature in K.

The direct reading type of instrument, although possibly less accurate than potentiometric is also used exclusively in modern soil laboratories. The e.m.f. of the glass electrode-calomel electrode cell is applied across a resistance, and the resulting current after amplification is passed through an ammeter causing deflection of the pointer across a scale marked in pH units. These instruments are available to operate on mains A.C. current. In most pH meters temperature control knob is provided to adjust at temperature of the test solution.

### 1.1.4 Electrode Potential Determination; Illustration with Calomel Electrode; Hydrogen Electrode; Standard Oxidation Potential

Generally, Nernst equation is used for the processes at an electrode to evaluate the single electrode potential. Let us consider that a zinc electrode is dipped in a solution of $Zn^{++}$ ions. Let the actual, process occurring at the electrode be one of the oxidation.

Electrode : $Zn/Zn^{++}$
Electrode process : $Zn \rightarrow Zn^{++} + 2e$

If $\xi_{Zn}$ and $\xi^\circ_{Zn}$ denote the electrode potentials of zinc in a solution of $Zn^{++}$ ions of activity $a_{Zn^{++}}$ and in a solution of $Zn^{++}$ ions of unit activity respectively, then by applying Nernst equation.

$$\xi_{Zn} = \xi^\circ_{Zn} - RT/2F \ln \frac{a_{Zn^{++}}}{a_{Zn}} \quad ...(1.1.4.1)$$

Since, activity of pure zinc metal, $a_{Zn} = 1$, we have

$$\xi_{Zn} = \xi^\circ_{Zn} - \frac{RT}{2F} \ln a_{Zn^{++}} \quad ...(1.1.4.2)$$

$\xi^\circ_{Zn}$ is the electrode potential of zinc in a standard solution of $Zn^{++}$ ions of unit activity. $\xi^\circ_{Zn}$ is called the standard electrode potential of zinc. Since it has been assumed, that oxidation occurs in the electrode, $\xi_{Zn}$ is really the oxidation potential of the electrode and $\xi^\circ_{Zn}$ is its standard oxidation potential. Hence in generalised form Nernst equation, where the potential of an electrode in which oxidation occurs may be expressed as

$$\xi_M = \xi^\circ_{M+n} - \frac{RT}{nF} \ln \frac{a_{oxidant}}{a_{reductant}} \quad ...(1.1.4.3)$$

where

R = universal gas constant = 8.32 joules per degree per mole,
T = absolute temperature,
F = Faraday = 96500 coulombs,
$a$ = activity.

In order to assign numerical values to the electrode potential it is necessary to choose a standard electrode and assign an arbitrary value to the potential of the same. For this purpose the reference electrode is the normal hydrogen electrode, (Pt) ½$H_2$ (1 atm) (gas) | $H^+$ ($a$ = 1) (electrode process: ½$H_2$ = $H^+ + e^-$) in which pure hydrogen gas at unit pressure is kept in contact with solution containing $H^+$ ion of unit activity through adsorption on Pt black by continuous bubbling of the gas. The potential of this normal hydrogen electrode is taken as zero at all temperatures. It should be emphasised that if the acid solution has $H^+$ ion activity other than unity, the electrode potential would no longer be zero for

$$\xi_{H_2} = \xi^{\circ}_{H_2} - \frac{RT}{nF} \ln a_{H^+} = -\frac{RT}{nF} \ln a_{H^+} \quad ...(1.1.4.4)$$

If $\quad a_{H^+} \neq 1, \xi_{H_2} \neq 0$

The potentials of other electrodes are expressed in reference to the normal hydrogen electrode. To evaluate the potential for any other single electrode, it is necessary to couple it with a standard or normal hydrogen electrode and the e.m.f. of the galvanic cell is measured potentiometrically. Since the e.m.f. of the cell is known and is equal to the algebraic sum of the two electrode potentials of which $\xi^{\circ}_{H_2} = 0$, the potential of the other electrode is obtained. If $\xi_x$ and $\xi^{\circ}_{H_2}$ are oxidation potentials of the electrode and the standard hydrogen electrode respectively, the e.m.f. (E) of the cell will be given as difference of the two *i.e.,*

$$E = \xi_x - \xi^{\circ}_{H_2} \quad ...(1.1.4.5)$$

If the given electrode functions as anode;

then $\quad E = \xi_{anode} - \xi_{cathode} = \xi_x - \xi^{\circ}_{H_2} = \xi_x$

But if the given electrode functions as cathode,

then $\quad E = \xi_{anode} - \xi_{cathode} = \xi^{\circ}_{H_2} - \xi_x = -\xi_x$

**Illustration: Determination of potential of calomel electrode**

The calomel electrode consists of mercury in contact with saturated solution of mercurous chloride and a large excess of potassium chloride solution which may be either saturated solution or normal solution.

Electrode: $Hg \mid Hg_2Cl_2\ (s) \mid Cl^-$ ;

Electrode process (oxidation): $2Hg^+ + 2Cl^- = Hg_2Cl_2$

When it is coupled with a standard $H_2$ – electrode, the calomel electrode functions as cathode.

The cell may be arranged as:

| Anode (–) | | | Cathode (+) |
|---|---|---|---|
| (Pt) $H_2$ (gas) (1 atm) | $H^+$<br>$a_{H^+} = 1$ | KCl sol$^n$<br>$Cl^-$ | $Hg_2Cl_2$ (s) – Hg |

Cell e.m.f. $\quad (E) = \xi^{\circ}_{H_2} - \xi_{cal}$ (where $\xi_{cal}$ = oxidation potential of calomel electrode) ...(1.1.4.6)

or $\quad E = 0 - \xi_{cal} = -\xi_{cal} \quad$ ...(1.1.4.7)

$$\text{Now} \quad \xi_{cal} = \xi^{\circ}_{cal} - \frac{RT}{F}\ln\frac{a_{Hg_2Cl_2}}{a_{Hg}a^2_{cl^-}} \qquad ...(1.1.4.8)$$

$$= \xi^{\circ}_{cal} - \frac{RT}{2F}\ln\frac{1}{a^2_{Cl^-}} (\text{since } a_{Hg}, a_{Hg_2Cl_2} \text{ are unity}) \qquad ...(1.1.4.9)$$

$$= \xi^{\circ}_{cal} - \frac{RT}{2F}(\ln 1 - 2\ln a_{Cl^-}) \qquad ...(1.1.4.10)$$

$$= \xi^{\circ}_{cal} - \frac{RT}{F}\ln a_{Cl^-} \qquad ...(1.1.4.11)$$

$$\therefore \quad E = -\left(\xi^{\circ}_{cal} + \frac{RT}{F}\ln a_{Cl^-}\right) \qquad ...(1.1.4.12)$$

Hence, at 25°C when $a_{Cl^-} = 1, \xi^{\circ}_{cal} = -E = -0.2680$ volt as experimentally determined.

**Hydrogen electrode**

Applying Nernst equation to hydrogen electrode already described;

$$\xi_{H_2} = \xi^{\circ}_{H_2} - \frac{RT}{F}\ln a_{H^+} = -\frac{RT}{F}\ln a_{H^+} (\text{since } \xi^{\circ}_{H_2} = 0) \qquad ...(1.1.4.13)$$

$$\text{or} \quad \xi_{H_2} = -2.303\frac{RT}{F}\log a_{H^+} \qquad ...(1.1.4.14)$$

$$\text{or} \quad \xi_{H_2} = -2.303\frac{RT}{F}\text{pH} \qquad ...(1.1.4.15)$$

Now the half cell 1.1.4.11 is coupled with a reference electrode, say a saturated calomel electrode, through a KCl bridge so that junction potential is eliminated.

If E is the measured e.m.f. of the cell, then,

$$E = \xi_{H_2} - \xi_{cal} \qquad ...(1.1.4.16)$$

$$= -\frac{RT}{F}\ln a_{H^+} - \xi_{cal}a_{H^+} \qquad ...(1.1.4.17)$$

$$= -\xi_{cal} + 2.303\frac{RT}{F}\text{pH} (\text{since} - \log a_{H^+} = \text{pH}) \qquad ...(1.1.4.18)$$

$$\textit{i.e.,} \quad \text{pH} = \left[\frac{F(E+\xi_{cal})}{2.303RT}\right] = \left[\frac{(E+\xi_{cal})}{0.059}\right] \qquad ...(1.1.4.19)$$

$$(\text{since,} \quad 2.303\frac{RT}{F} = 0.059, \text{ at } 25°C).$$

$$\text{or} \quad \text{pH} = \left(\frac{E-0.268}{0.059}\right) \qquad ...(1.1.4.20)$$

### 1.1.5 Potentiometric Method

A metal is regarded as an assembly of metal ions of free electrons. When the metal is in contact with water, some metal ions enter into the liquid due to a tendency in the metal, called by Nernst as 'electrolytic solution tension'. As some metal ions leave the solid, the solid becomes negatively charged and the solution positively charged. In consequence, due to electrostatic force, any further transference of the metal ions is prevented and the ions attracted by the negatively charged metal, remain near the metal surface forming a double layer. If the metal is placed in a solution containing its own ions, the metal ions from the solution in virtue of their osmotic pressure may enter into the metal rendering its surface positively charged. Again by attraction, the anions would flock near the positively charged surface and forms a double layer. There is thus always a double layer at the contact of electrode metal and electrolyte. Hence, a difference of potential exist between metal phase and solution phase. This potential difference in the half cell is called the single electrode potential. In this context it may be stated, that a galvanic cell, a device in which free energy of a chemical process is converted to electrical energy, must necessarily consist of two electrodes; positive and negative. Each of these two is known as a half cell or single electrode. The process occurring in the cell, ultimately causes transfer of electrons from the electrode to the electrolyte and vice versa, resulting into a flow of current. The cell e.m.f. is given by the algebraic sum of its electrode potentials. Therefore,

$$\text{e.m.f. (E)} = \xi^{\text{oxd}}_{\text{anode}} + \xi^{\text{red}}_{\text{cathode}} \quad \text{or} \quad \text{(E)} = \xi^{\text{oxd}}_{\text{anode}} - \xi^{\text{oxd}}_{\text{cathode}} \quad \text{...(1.1.5.1)}$$

where

$\xi^{\text{oxd}}_{\text{anode}}$ = oxidation potential of anode,

$\xi^{\text{red}}_{\text{cathode}}$ = reduction potential of cathode.

It is to be remembered that reduction potential of an electrode is same as its oxidation potential with the sign changed. Usually anode of a cell is written in the left and cathode in the right. It is also a common convention that current in external circuit flows from cathode to anode although the electrons are flowing in the opposite direction through the wire.

### 1.1.6 Liquid Junction Potential

The liquid junction potential is the most important source of error when using the glass electrode calomel electrode system. When two solutions of different strength or composition come into contact the more concentrated solution will diffuse into the more dilute. If the ions of the diffusing solution move at different speeds the dilute solution will assume an electric charge with respect to the concentrated solution corresponding to that of the faster moving ion. For example, if the diffusing anions move more quickly than the cations they will cause the dilute solution to become negative with respect to the concentrated solution. The resulting difference in potential across the interface of the solutions is called the **liquid junction potential** ($E_J$) and adds to or subtracts from the electric potential. Such a potential is likely to arise at the liquid junction between a soil suspension, and the salt bridge of the calomel electrode. The presence of colloids or suspensions has a marked effect on, liquid junction potentials and hence this error may be more important, in soil pH measurements than when using pure solutions. Attempts have been made to allow for liquid junction potentials by calculation. The calculation involves knowledge of activity coefficients and even for true solutions have proved, to be of little use and would be quite impossible to derive for soil suspensions. One procedure to minimize the liquid junction potential is to use saturated potassium chloride solution as the salt bridge. It is the relative mobilities of the oppositely charged ions at the interface that decide the potential gradient and thus it is desirable to equate these mobilities as far as possible. Potassium chloride is used as potassium ions and chloride ions have about the same mobility, and if the concentration of the salt is much greater than that of other electrolytes present it will be responsible for transferring almost, all the current across the liquid junction.

### 1.1.7 Drifting of Soil pH

Occasionally a soil exhibits pH drift that is the pH will slowly but continuously increase or decrease, and it is difficult to decide upon the true value. There is no hard and fast rule for dealing with this problem. Some workers recommend allowing the soil paste to stand, for an arbitrary period of time say 15 minutes with the electrodes in place and the instrument on and to accept the reading obtained. Whatever is done, it is obvious that a single figure will have little significance and it is best to record, that the pH is drifting and to give the limits over a certain period of time. The most important result of the measurement is that the pH does drift and in which direction.

### 1.1.8 Experimental Determination of Cell E.M.F.

The e.m.f. of a cell is measured with the help of a potentiometer. The principle involved can be clearly understood from Fig. 1.1.

AB is the potentiometer slide wire of a uniform cross-section and having a high resistance. A storage cell C is connected, across the terminals of the slide wire AB, such that potential drops from A to B. Now the cell X whose e.m.f. is required is connected to A so that its e.m.f. opposes that of C. (*i.e.,* A is connected to positive terminals of both X and C). The other terminal of cell X is connected through a galvanometer (G) to a sliding contact P. This is moved along the slide wire until, there is no deflection in the galvanometer.

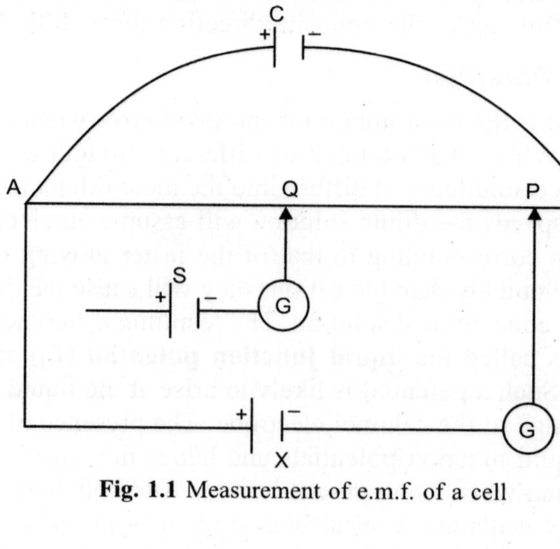


**Fig. 1.1** Measurement of e.m.f. of a cell

This means that the emf of cell X just balances the drop of potential between A and P. Next a standard cell (S) taken to replace the cell X and the experiment is repeated. The e.m.f. of the cell S now opposes that of C in the slide wire. Let the contact point, now be Q when there would not be any deflection in the galvanometer. This means the drop of potential in the slide wire from A to Q just balances the emf of the standard cell.

If the $E_x$ and $E_s$ be the e.m.f. of the given cell and standard cell then

$$\frac{E_x}{E_s} = \frac{\text{drop of potential from A to P}}{\text{drop of potential from A to Q}} = \frac{\text{resistance of AP}}{\text{resistance of AQ}} = \frac{\text{length AOP}}{\text{length AQ}}; \qquad \text{...(1.1.8.1)}$$

Since, the wire is of uniform cross-section, the two lengths being known and since $E_s$, (the e.m.f. of the standard cell) is known, $E_x$ can easily be determined.

### 1.1.9 Care and Maintenance

The most delicate part of the pH-meter is the glass electrode which may crack or break, if handed roughly or may dry up when left out of water for a long period. Under such situation the electrode should be immersed in 0.1(N) HCl and then in distilled water for 24 hours or more and checked again. The pH-meter is switched on and 10–15 minutes time is allowed for warming up.

## 1.2 ELECTRICAL CONDUCTANCE: GENERAL DISCUSSION

### 1.2.1 Ohm's Law (Resistance, Specific Resistance, Conductance, Equivalent Conductance)

Ohm's law states that temperature and other physical conditions remaining constant, the current flowing through a conductor is directly proportional to the potential difference between both ends of the conductor.

Let $V_A$ and $V_B$ be the potentials at the ends A and B respectively of a conductor AB (Fig. 1.2.)

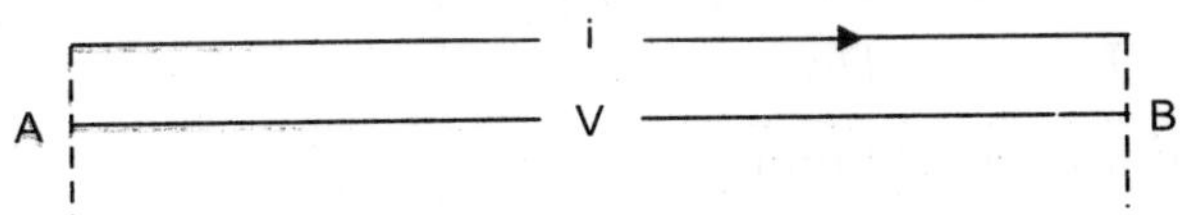


**Fig. 1.2** Ohm's law

---

The operational definition of the pH of a solution X is that it is given by

pH (X) = pH(s) + E/2.303 RT/F where E is the e.m.f. of the cell,

Pt|$H_2$ |X(aq.)3.5 M KCl (aq.) |S(aq.)| $H_2$ | Pt;

the solution S being a solution of standard pH. The primary standard is a 0.05 (M) aqueous solution of pure potassium hydrogen phthalate, of which the pH is defined as being exactly 4 at 15 °C and at other temperatures ($t$ °C) as pH (S) = $\left[4+\sqrt{\frac{(t-15)}{2}\times 10^4}\right]$, if $t$ lies between 0 and 55 °C (e.g., 4.005 at 25 °C).

The values of pH given by this definition differ very slightly from the formal definition.

Let $i$ be the current flowing through AB, then according to Ohm's law

$$i \propto (V_A - V_B) \text{ or } \frac{V_A - V_B}{i} = R \text{ (a constant)} \quad ...(1.2.1)$$

*i.e.*, $V/i = R$ where $V_A - V_B = V$ (say) ...(1.2.2)

Equation (1.2.2) can be written as $V = iR$ ...(1.2.3)

and $i = V/R$ ...(1.2.4)

Equations (1.2.2), (1.2.3) and (1.2.4) are known as mathematical form of Ohm's law. The proportionality constant (R) is called the resistance of the conductor, the value of which depends on the materials and dimension of the conductor.

From equation (1.2.4) it is evident that for the same potential difference applied across a conductor, an increase in the resistance of the conductor lowers the current through it and vice versa. Thus the resistance of a conductor may be defined, as that property of the conductor by virtue of which, it opposes the flow of electricity through it. It is expressed quantitatively as the ratio of the potential difference across the conductor and the current flowing through it. The practical unit of resistance is ohm generally expressed by the symbol (Ω) omega. The resistance of conductor is 1 ohm if the current flowing through it is 1 ampere when the potential difference between its ends is 1 volt. Thus volt/ampere = ohm.

In a metallic conductor of length ($l$) cross section ($a$) the resistance (R) is given by

$$R = \rho \frac{l}{a} \quad ...(1.2.5)$$

where ρ is the specific resistance or resistivity. It is the resistance of unit length of the conductor of unit cross-section.

The reciprocal of resistance is termed as a conductance (∧) and the reciprocal of resistivity is the specific conductance of conductivity (L) or (κ)

Hence, conductivity L or $\kappa = 1/\rho$ ...(1.2.6)

The conductance of a given solution,

$$\wedge = \frac{1}{R} = \frac{1}{\rho} \cdot \frac{a}{l} = \frac{aL}{\rho} \quad ...(1.2.7)$$

Therefore,

$$L = \frac{1}{R} \cdot \frac{l}{a} \quad ...(1.2.8)$$

The resistance is expressed in units of ohm (Ω) and the conductance has units of reciprocal ohm or mho.

Now from equation (1.2.5) if $l = 1$, $a = 1$,

the specific conductance or conductivity L or (κ) = ∧ (conductance) ...(1.2.9)

Therefore, specific conductance or conductivity can be defined as the conductance of a solution enclosed between two electrodes of 1 sq. cm. area and 1 cm apart.

The conductance of the solution depends upon the number of ions present and hence on the concentration. To compare the conductivity of different solutions it is necessary to take the concentration of the solutions into consideration. It is done by using equivalent conductance ($\lambda$).

The equivalent conductance is defined as the conductance of a solution containing 1g equivalent of the dissolved electrolyte such that the entire solution is placed between two electrodes 1 cm apart. As direct determination of the quantity would need electrodes of enormous sizes, the equivalent conductance ($\lambda$) is always evaluated through measurement of specific conductance or conductivity with the help of equation (1.2.8).

Let the solution of the electrolyte has a concentration of C g equivalent per litre then the volume of the solution containing 1 g equivalent would be 1000/C cubic centimetre. If this volume is imagined to be placed between two electrodes 1 cm apart , ($l$ = 1), the cross section of the column of solution or electrodes would be 1000/C sq. cm. Hence equivalent conductance of the solution would be,

$$\lambda = a/l \cdot L = 1000/C \times 1 \times L = 1000\ L/C \qquad ...(1.2.10)$$

L or $\kappa$ being the specific conductance or conductivity.

An alternative unit, called molar conductance (m) is defined as the conductance of a solution containing 1 g mole per litre, the solution being placed between two electrodes 1 cm apart.

Hence $\mu = 1000\ \kappa\ C'$, is the molar concentration ...(1.2.11)

### 1.2.2 Measurement of Conductivity

The specific conductance (L or $\kappa$ ) or conductivity of a solution is always obtained by measuring the resistance (R) of the solution taken in a suitable container of known dimensions called conductivity cell, the cell constant of which has been determined by calibration with a solution of accurately known conductivity e.g., a standard KCl solution. The instrument used for electrical conductivity measurement is known as conductivity bridge. A typical system consists of an alternating current (A.C.) Wheatstone bridge, a primary element of conductivity cell and a null balance indicator (as in 'solubridge') or an electronic eye as in the conductivity meter.

The passage of a current through a solution of an electrolyte may produce changes in the composition of the solution in the vicinity of the electrodes; the potentials may thus arise at the electrodes with the consequent introduction of serious errors in the conductivity measurements unless such polarisation effects can be reduced to negligible proportions. These difficulties are generally overcome by the use of alternating currents for the measurements so that the extent of electrolysis and the polarisation effects are greatly reduced.

Generally conductivity cells are constructed of Pyrex or other resistance glass and fitted with platinised platinum electrodes, the platinising also helps to minimise the polarisation effects. The distance '$l$' between two electrodes in a cell is fixed For most purposes good results are obtained by clamping a commercially available 'dip cell' inside a beaker containing the test solution. The solutions obey Ohm's law. The cell is placed in one arm of a Wheatstone bridge circuit and the resistance measured.

### 1.2.3 Wheatstone Bridge Principle

In the year 1843, Charles Wheatstone, the first Professor of Physics at King's College, London, invented one of the most accurate and commonly used methods of measuring resistance. It is known as Wheatstone bridge method. By this method the ratio of two resistances is determined and if the value of one of them is known, the value of the other is obtained (Fig. 1.3) shows the circuit diagram of Wheatstone bridge.

Four resistances P, Q, R and S are connected to form a close network ABCD. A galvanometer G is connected between the junctions B of P and Q and D of R and S. A cell E is connected P, Q, Rod S between the other two junctions viz. A of P and R and C of Q and S. AB, BC, AD and AC are called the 1st, 2nd, 3rd and 4th arm of the bridge respectively. AB and BC are also called the ratio arms. By properly adjusting the value of the resistances, the current through the galvanometer may be reduced to zero. This happens when point B and D are maintained at the same potential (Fig. 1.3). The galvanometer then shows no deflection and the network is said to be balanced. It can be shown that the resistances in the four arms of the bridge then satisfy the relation

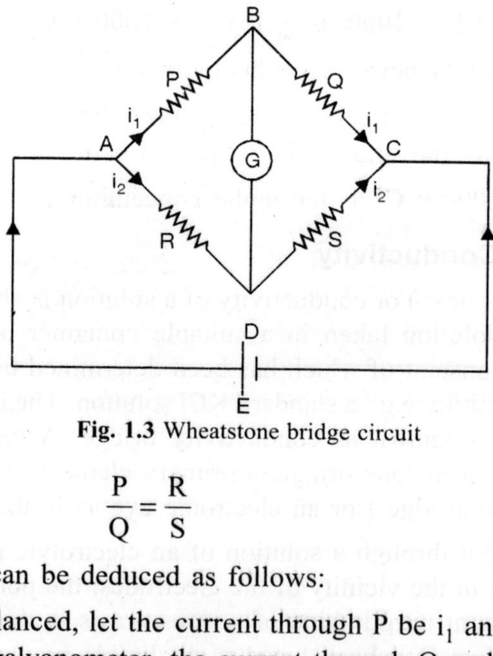


**Fig. 1.3** Wheatstone bridge circuit

$$\frac{P}{Q} = \frac{R}{S} \qquad \text{...(1.2.3.1)}$$

The equation 1.2.3.1 can be deduced as follows:

When the bridge is balanced, let the current through P be $i_1$ and through R be $i_2$. Since no current flows through the galvanometer, the current through Q and S must also be equal to $i_1$ and $i_2$ respectively. Moreover, the potentials at B and D are equal

i.e., $$V_B = V_D \qquad \text{...(1.2.3.2)}$$

If $V_A$ and $V_C$ be the potentials at A and C respectively, then

$$V_A - V_B = V_A - V_D \qquad \text{...(1.2.3.3)}$$

or $$i_1 P = i_2 R \qquad \text{(by Ohm's law)} \qquad \text{...(1.2.3.4)}$$

Again $$V_B - V_C = V_D - V_C \qquad \text{...(1.2.3.5)}$$

or $$i_2 Q = i_2 S \qquad ...(1.2.3.6)$$

Dividing (1.2.3.4) and (1.2.3.6), we get

$$\frac{i_2 P}{i_1 Q} = \frac{i_2 R}{i_2 S} \qquad ...(1.2.3.7)$$

Hence $$\frac{P}{Q} = \frac{R}{S} \qquad ...(1.2.3.8)$$

Therefore, $$R = \frac{P}{Q} \cdot S \qquad ...(1.2.3.9)$$

Hence, if the value of R be unknown, it can be found from a knowledge of S and the ratio P/Q. Since the method requires 'no deflection' of the galvanometer it is known as the null method. The balance condition may be written as $\frac{Q}{S} = \frac{P}{R}$. This shows that the balance condition remains the same if the positions of the galvanometer and the battery be interchanged. The branches AC and BD are therefore said to be conjugated to each other. It is obvious that the balance condition is independent of the current supplied by the cell, the resistance of the galvanometer, the internal resistance of the cell and the resistance connected in series with the galvanometer and the battery.

In experimental arrangement (Fig. 1.4) the cell X is connected to one arm of the bridge, the other arm QD carries a variable resistance $R_3$. PQ is an uniform slide wire on which moves a contact point C. The contact, point C is connected through a ear-phone to a point D junction of the other two arms PD and QD containing the cell and the variable resistance $R_3$. An A.C. current is used in the circuit otherwise electrolysis would occur and the concentration would change. The temperature is controlled thermostatically. The current from the source enters at P and Q and divides into two parallel branches along PCQ and PDQ. Using a definite resistance $R_3$ in the arm DQ, the contact point C is moved along the slide wire until no sound is produced in the ear phone *i.e.*, until no current passes along DC. Under this condition potentials and C and D are the same.

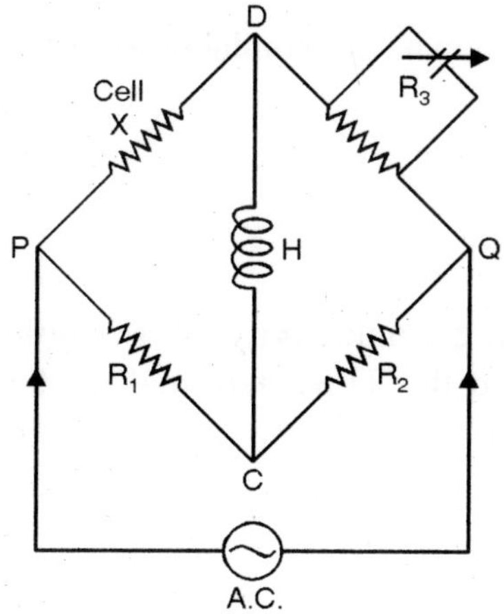


**Fig. 1.4** Conductivity determination circuit

Hence, $$\frac{X}{R_3} = \frac{R_1}{R_2} \text{ or } X = \frac{R_1}{R_2} \cdot R_3 = \frac{l_1}{l_2} \cdot R_3 \qquad ...(1.2.3.10)$$

where X is the resistance of the solution, $R_1$ and $R_2$ are the resistances of the solution of the two portions of the slide wire, the ratio arms $l_1$ and $l_2$. In fact, the $\frac{R_1}{R_2}$ is the ratio of lengths CP/CQ, when a wire of uniform cross-section is used. The resistance of the solution X *i.e.,* of the cell, is thus known. Theoretically when balance point is reached by moving the contact point C, there should be no sound in the earphone but due to capacitance arising from the cell, some little sound occurs at the balance point. The point where the sound is minimum is taken as the balance point. By inserting a variable condenser parallel to the standard resistance $R_3$, the capacitance effect of the conductivity cell can be eliminated to a large extent and much improved balancing is possible.

To know the conductivity *i.e.,* specific resistance it would be necessary to determine the cross section and the distance between the electrodes of the cell used. The ratio ($l/a$) known as 'cell constant' (K) is determined in an alternative way. Using conductivity cells of accurately known dimensions ($l$ and $a$) Kohlrausch and his co-workers determined very precisely the specific conductance of standard solutions of pure KCl at different temperatures. In order to ascertain the cell constant ($l/a$) of a conductivity cell used in the laboratory, the resistance of KCl solution of 0.1 or 0.01 molar strength is measured. Let the resistance of the KCl solution is found to be $r$. From equation 1.2.8 the cell constant $K = l/a = L_s \cdot r$ ...(1.2.3.11)

where $L_s$ is the conductivity of KCl solution known from table value (Appendix VIII). The cell constant of a particular cell is thus known. For a given solution the resistance (R) is measured in usual way with the Wheatstone bridge circuit. The specific conductance or conductivity (L) of the solution

$$L = \frac{l}{a} \cdot \frac{1}{R} = \frac{K}{R} \qquad (1.2.3.12)$$

Since K and R both are known, the conductivity of the given solution is also known the equivalent conductance.

Equivalent conductance

$$(\lambda) = 1000\frac{l}{C} \qquad ...(1.2.3.13)$$

Practically while measuring conductivity of a solution a 'dip cell' is supported in the solution, and then connected to the TEST terminal of the conductivity bridge. The selector switch is set to the appropriate conductance range, and the dial is rotated until a balance is indicated on the magic eye. The conductivity may be calculated by multiplying the observed conductance by the cell constant.

### 1.2.4 Types of Conductivity Meters

Cambridge conductivity meter (bridge) is a mains (A.C.) operated Wheatstone bridge; there is a built in 1000 cycles per second oscillator. This instrument is supplied by Cambridge Instrument Co. Ltd., Grosvenor Place, London, U.K. Messers ELICO (India) Pvt. Ltd. has also developed a conductivity bridge (50 c/s to 1000 c/s) which has a similar type of 'magic eye' detector as in the case of solubridge. M/s Systronics and other manufacturers has also come out with similar products.

### 1.2.5 Care and Maintenance

The conductivity meter has a long life and it rarely goes out of order. If it does, the metallic cover may be unscrewed and examined for loose contact in the internal wiring or the vacuum tube may be checked. Often the trouble arises from the conductivity cell. The essential component of the cell is the two electrodes coated with platinum black and rigidly set at a specific distance (5 mm or so). Sometimes due to inadequate washing, a clay film is deposited on the electrodes. It can be removed by repeated washings with distilled water. In case the cell needs drastic cleaning then freshly prepared chromic-sulphuric acid which is always quite warm is used and the cell finally washed several times with distilled water. The chromic acid, must not be allowed to get in contact with the rubber bulb of the conductivity cell or any metallic parts.

## 1.3 COLORIMETRY AND SPECTROPHOTOMETRY—GENERAL DISCUSSION AND THEORETICAL CONSIDERATION

The variation of the colour of a system with change in concentration of some component forms the basis of colorimetric analysis. The colour develops due to the formation of a coloured compound by the addition of an appropriate reagent, or it may be inherant in the desired constituent itself. The intensity of colour is then compared with that obtained by treating a known amount of the substance in the similar manner. Colorimetry is thus the determination of the concentration of a substance by measurement of the relative absorption of light with respect to a known concentration of the substance. In visual colorimetry natural or artificial, white light, is generally used as a light source and determinations are normally done with a simple instrument termed as a colorimeter. When the eye is replaced by a photoelectric cell, thereby largely eliminating the errors due to the personal characteristics of each observer, the instrument is termed as photoelectric colorimeter. The latter is usually used with the light contained within a comparatively narrow range of wavelength furnished by passing white light through filters *i.e.,* materials in the form of the plates of coloured glass, gelatin etc. transmitting only a limited spectral region; the name filter photometer is sometimes applied to such instrument.

In spectrometric analysis a radiation source is used which extend into the ultraviolet region of the spectrum. From this, definite wavelength of radiation are chosen possesing a bandwidth of less than 1 nm. This process necessitates the use of more complicated and consequently more expensive instrument. The instrument employed for this purpose is a spectrophotometer which is really two instruments in one cabinet, a spectrometer and a photometer. An optical spectrometer is an instrument, possessing an optical system which can produce dispersion of incident electromagnetic radiation, and with which measurements can be made of the quantity of transmitted radiation at selected wavelengths of the spectral range. A photometer is a device for measuring the intensity of transmitted radiation. When combined in the spectrophotometer, the spectrometer and the photometer are employed cojointly to produce a signal corresponding to the difference between the transmitted radiation of reference material and that of a sample at selected wavelengths. The most important advantage of spectrophotometric analysis is that they provide a simple means for determining minute quantities of substances.

When light is passed through a given liquid or solution the absorption does not occur at all wavelengths. At a particular wavelength or within a small range of the same light is considerably absorbed. The decrease in intensity of incident radiation during its passage through the absorbing medium is governed by two laws: Lambert's law and Beer's law. In the combined form they are referred to as the Beer-Lambert law.

### 1.3.1 Beer-Lambert's Law

This law states that when a monochromatic light passes through a transparent medium, the rate of decrease in intensity with the thickness of the absorbing medium is proportional to the intensity of the penetrating radiation. Let us consider a thin layer of the medium of thickness $dl$ and let I be the intensity of the radiation entering it, then **Lambert's law** can be expressed by the differential equation as:

$$-\frac{dI}{dl} = kI \qquad \text{...(1.3.1.1)}$$

or

$$\int_{I_0}^{I} \frac{dI}{I} = -k\int_0^l dl \qquad \text{...(1.3.1.2)}$$

or

$$\ln \frac{I}{I_0} = -kl \qquad \text{...(1.3.1.3)}$$

or

$$I = I_0\, e^{-kl} \qquad \text{...(1.3.1.4)}$$

where, $I_0$ is the intensity at $l = 0$, and I, the intensity at distance $l$. The proportionality constant $k$ is called the absorption coefficient of the substance.

By changing from natural to common logarithms the equation (1.3.1.4) can also be written as

$$I = I_0\, 10^{-al} \qquad \text{...(1.3.1.5)}$$

where

$a = k/2.3026 = 0.4343k$ and is termed as 'extinction coefficient'.

The **extinction coefficient** is generally defined as the reciprocal of the thickness (in cm) required to reduce the light by 1/10 of its intensity. It is obvious that the proportion of the amount of light, absorbed $(I_0 - I)/I_0$ with equal thickness ($l$) of the absorbing material will be the same and this proportion is independant of the intensity of incident light.

When the absorbing substance is present in solution, the absorption of light also depends upon the concentration **Beer's law** states that the rate of decrease in intensity of radiation absorbed is proportional to the intensity of radiation and to the concentration of the solute. Mathematically

$$\frac{dI}{dl} = -k'cI \text{ (where } c = \text{concentration)} \quad ...(1.3.1.6)$$

or $$\int_{I_0}^{I} \frac{dI}{I} = \int_{o}^{l} k' c\,dl \quad ...(1.3.1.7)$$

$$\ln \frac{I}{I_0} = -k' cl \quad ...(1.3.1.8)$$

$$\frac{I}{I_0} = e^{-k' cl} \quad ...(1.3.1.9)$$

Therefore, $$I = I_0 \cdot e^{-k' cl}$$

Rewriting equation (1.3.1.8)

$$2.303 \log_{10} \frac{I}{I_0} = -k' cl \quad ...(1.3.1.10)$$

or, $$\log_{10} \frac{I}{I_0} = -0.4343\ k' cl \quad ...(1.3.1.11)$$

or, $$\log_{10} \frac{I_0}{I} = 0.4343\ k' cl \quad ...(1.3.1.12)$$

or, $$\log_{10} I_o/I = \varepsilon cl \quad ...(1.3.1.13)$$

or, $$I = I_0\, 10^{-\varepsilon cl} \quad ...(1.3.1.14)$$

where $\varepsilon$ is called the molar extinction coefficient such that $\varepsilon = 0.4343\ k'$. The value of $\varepsilon$ is specific for a given substance for a given wavelength of light. Equation (1.3.1.8) is the fundamental equation of colorimetry and spectrophotometry and is often spoken of as the **Beer-Lambert law**.

The quantity $\log_{10} \frac{I_0}{I}$ is generally called the optical density (O.D) or absorbancy so that

$$O.D = \log_{10} \frac{I_0}{I} = \varepsilon cl \quad ...(1.3.1.15)$$

when log $(I_0/I)$ is plotted against concentration of solution taken in a column of definite thickness, a straight line is obtained. The slope of the line gives the value of molar extinction coefficient.

It will be apparent that there is a relationship between the absorbance(A) the transmittance (T) and the molar extinction coefficient ($\in$) since,

Absorbance (A) or Optical density (O.D)

$$= \varepsilon cl = \log \frac{I_0}{I} = \log \frac{1}{T} = -\log T \quad ...(1.3.1.16)$$

The scales of spectrophotometers are often calibrated, to read directly in absorbances and frequently also in percent transmittance.

For matched cells (*i.e.*, $l$ = constant) the Beer-Lambert law may be written as:

$$c \propto \log_{10} I_0/I \quad ...(1.3.1.17)$$

*i.e.*,

$$c \propto O.D \quad ...(1.3.1.18)$$

Hence by plotting O.D (or log 1/T), as ordinate, versus concentration as abscissa, a straight line will be obtained and this will pass through the point C = O, A = O (T = 100%). This calibration line may then be used to determine unknown concentrations of solutions of the same material after measurement of absorbances.

### 1.3.2 Deviation from Beer's Law

Beer's law generally holds good over a wide range of concentration if the structure of the coloured non electrolyte in the dissolved state does not change with concentration. Small amount of electrolytes, which do not react chemically with the coloured components, do not usually affect the light absorption, large amounts of electrolytes may result in a shift of the maximum absorption and may also change the value of extinction coefficient. Discrepancies are normally observed when the coloured solute ionises, dissociates or associates in solution as because the nature of the species in solution will vary with the concentration. The law also fails if the coloured solute forms complexes, the composition of which depends upon the concentration. Also discrepancies may occur when monochromatic light is not used. The plot of log ($I_0$/ I) versus concentration must be a straight line passing through the origin which indicates conformity to the law.

### 1.3.3 Spectrophotometer: Instrumentation

Spectrophotometer from stand point of analytical chemistry are those instruments which enable one to measure absorbance (or, transmittance) at various wavelengths. A spectrophotometer may also be regarded as a refined filter photoelectric photometer which permits the use of continuously variable and more nearly monochromatic bands of light. The essential, parts of a spectrophotometer are (*i*) a source of radiant energy, (*ii*) a monochromator (filter, prism or diffraction grating) *i.e.*, a device for isolating monochromatic light *i.e.*, light of a single frequency or more precisely expressed narrow bands of radiant energy from the light source (*iii*) glass or silica cells for the solvent and for the solution under test and (*iv*) a device to receive or measure the beam or beams of radiant energy passing through the solvent or solution in terms of electricity generated. Generally tungsten filament lamp and hydrogen discharge are used as light source, the former for measurements down to 320 nm and the latter for the measurements in the UV region below 360 nm (Fig. 1.5).

Most modern ultraviolet/visible spectrophotometers are double beam instruments which generally covers the range between about 200 nm and 800 nm. In these instruments the monochromated beam of radiation, from tungsten and deuterium lamp sources is divided into two identical beams of equal intensity, one of which passes through the reference cell and other through the sample cell.

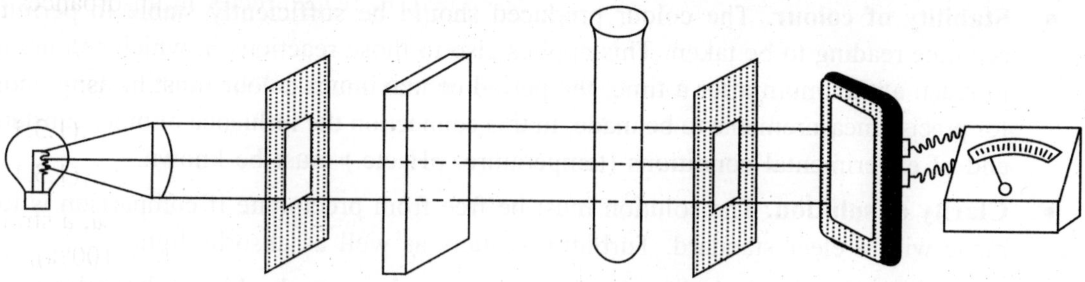


**Fig. 1.5** Components of optical photometers and spectrometers

Dispersion grating can be employed to obtain monochromatic beam of light from polychromatic radiation(UV-VIS). As the dispersion of a single beam or grating is very small, it is not possible to isolate very narrow bandwidths. Thus, light from the first dispersion is passed through a slit and then send to the second exit slit. The main advantage of the second dispersion is that the bandwidth of the emergent light, increase and the light passing through the exit slit is almost monochromatic. Also most of the stray light is suppressed.

The signal for the absorption of contents of the reference cell is automatically electronically subtracted from that of the sample cell giving a net signal corresponding to the absorption for the components in the sample solution. The instruments also possess digital display for the instantaneous reading of the absorbance values as these are measured.

When the sample absorbs light, its intensity is lowered. Thus the photo electronic cells will receive an intense beam from the reference cell and a weak beam from the sample cell. This results in the generation of pulsating or alternating currents, which flow from the photoelectric cells to the electronic amplifier. The amplifier is coupled to a small servomotors which drives an optical wedge into the reference beam until the photoelectric cell receive light of equal intensities from the sample as well as the reference beams.

Colorimetric method will often give more accurate results at low concentrations than the corresponding titrimetric or gravimetric methods. The criteria for a satisfactory colorimetric analysis are:

- **Specificity of colour reaction.** Very few reactions are specific for a particular substance, but many give colours for a small group of related substances only *i.e.,* are selective. By utilising such devices as the introduction of other complex forming compounds, by altering the oxidation states and control of pH, close approximation to specificity may be obtained.

- **Proportionality between colour and concentration.** For visual colorimeters it is important that the colour intensity should increase linearly with the concentration of the substance to be determined.

- **Stability of colour.** The colour produced should be sufficiently stable to permit an accurate reading to be taken. This applies also to those reactions in which colours tend to reach a maximum after a time; the period of maximum colour must be long enough for precise measurements to be made. In this connection the influence of other substances and of experimental conditions (temperature, pH etc.) must be known.
- **Clarity of solution.** The solution must be free from precipitate if comparison is to be made with a clear standard. Turbidity scatters as well as absorbs light.
- **Reproducibility.** The colorimetric procedure must give reproducible results under specific experimental conditions.
- **High sensitivity.** It is desirable, particularly when minute amount of substances are to be determined, that the colour reaction be highly sensitive. It is also desirable that the reaction product absorb strongly in the visible rather than in the ultra-violet; the interfering effect of other substances in the ultra-violet is more pronounced.

In view of selective character of many colorimetric reactions, it is important to control the operational procedure so that the colour is specific for the component being determined. Use may be made of the following processes in order to render colour reactions specific and/or to separate the individual substances:

- Suppression of the action of interfering substances by the formation of complex ions or of non-reactive complexes.
- Adjustment of the pH; many reactions take place within well defined limits of pH.
- Removal of interfering substances by extraction with an organic solvent, sometimes after suitable chemical treatment.
- Application of physical methods utilising selective absorption chromatographic separations and ion exchange separations.

### 1.3.4 Standard Curve

The usual method of use of spectrophotometer requires the construction of standard curve (also termed as reference or calibration curve) for the constituent being determined. Suitable quantities of the constituent are taken and treated in the same way as the sample solution for the development of colour and the measurement of the transmittance (or absorbance) at the specified wavelength. The absorbance log $(I_0/I)$ is plotted against concentration: a straight line plot is obtained if Beer's law is obeyed. When the absorbance is directly proportional to the concentration only a few points are required to establish the line; when the relationship is not linear a greater number of points will generally be necessary. The standard curve should be checked at intervals. When plotting the standard curve it is customary to assign a transmission of 100% to the blank solution (reagent solution plus double distilled water); this represents zero concentration of the constituent. The readings are continued with a series of standard solutions and then with test solutions. A calibration curve is drawn relating the concentration of the standards to the absorbance values, using the relations

$$\%T = \frac{I}{I_0} \times 100 \qquad \ldots(1.3.4.1)$$

where T = transmittance

Thus
$$\log(\%T) = \log 100 + \log \frac{I}{I_0} = 2 - \log \frac{I_0}{I}; \qquad \ldots(1.3.4.2)$$

and the concentrations of the test solutions are obtained from corresponding absorbance values.

It may be mentioned that some colour solution have appreciable temperature coefficient of transmission, and the temperature of determination should not differ appreciably from that at which calibration curve was prepared.

## 1.4 FLAME SPECTROMETRY—GENERAL DISCUSSION AND ELEMENTARY THEORY

## RELEVANT BACKGROUND INFORMATION

### 1.4.1 Electromagnetic Radiation

Light and its various properties present some of the most important phenomena in the whole realm of physics and chemistry. All the properties of light can be explained by two complimentary theories; the corpuscular theory and the wave theory. Various phenomenon viz., interference, polarization, diffraction etc., are very well explained, considering wave nature of light. However, some effect like photoelectric effect, Compton effect are well described considering the particle nature of light. Light therefore, exhibits dual nature. Recent advances in modern physics postulates: when examined on an atomic scale the concept of particle and wave melt together; particles taking on the characteristics of waves and waves the characteristics of particles. Like light there are various forms of electromagnetic radiations such as ultra-violet, infra-red, X-rays, radio-waves etc. Some of the important characteristics of electromagnetic radiation are:

- These are produced by the oscillation of electric charge and magnetic field residing on the atom. The electric and magnetic components are mutually perpendicular to each other and are coplanar.
- These are characterised by their wavelengths, frequencies or wave numbers.
- The energy carried by an electromagnetic radiation is directly proportional to its frequency. The emission or absorption of radiation is quantised and each quantum of radiation is called a photon.
- When visible light is passed through a prism, it is split up into seven colours VIBGYOR which corresponds to definite wavelengths.

### 1.4.2 Electromagnetic Spectrum

The arrangement of all types of electromagnetic radiations in order of their increasing wavelengths or decreasing frequencies is known as complete electromagnetic spectrum. The radiations having wavelengths in the range of 3800 Å – 7600Å are known as visible radiation since human eye can detect only these radiations.

The complete range of electromagnetic spectrum is furnished in Fig. 1.6.

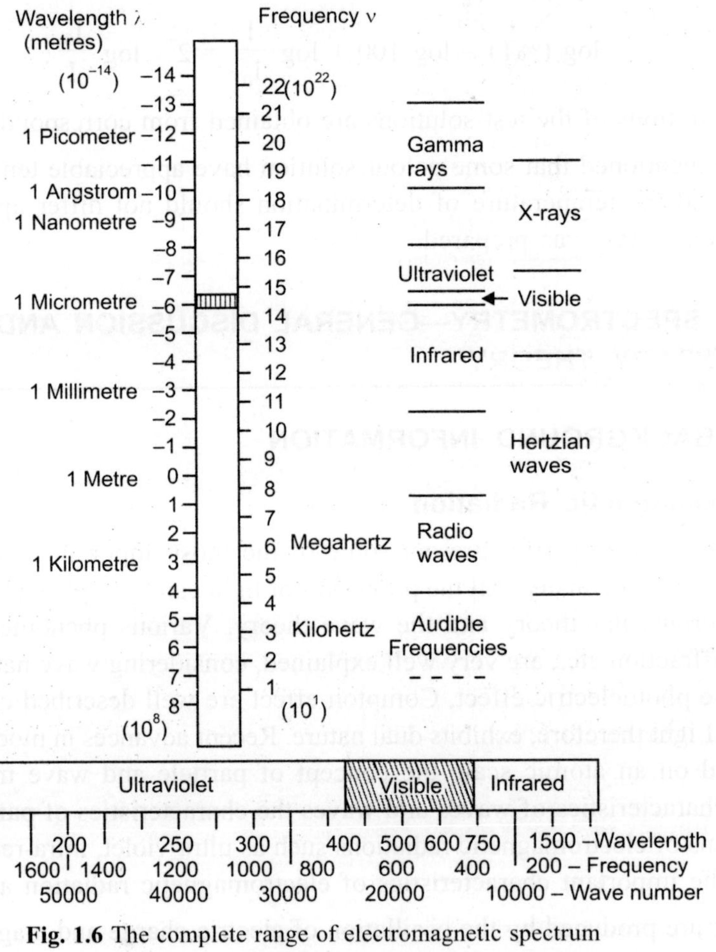


**Fig. 1.6** The complete range of electromagnetic spectrum

### 1.4.3 Wave Nature of Light

According to the wave theory, light travels in the form of waves. A wave is a sort of disturbance which originates from the vibrating sources. It travels in continuous sequence of alternating crests and troughs. The waves travel through space, at right angles to the vibratory motion of the object. Waves of visible light and those of other energy radiations are characterised by the following properties:

**Wavelength.** It is the distance between the two adjacent crests or troughs in a particular wave. It is denoted by the letter λ (Lambda). It is expressed in Angstrom (Å) units or nanometer (nm). Visible light, constitutes waves ranging from 3800 Å (violet end) to 7600Å (red end). Different colours of light have different values of their wavelength.

1Å = $10^{-8}$ cm    1 nm = $10^{-7}$ Å = 1 mm

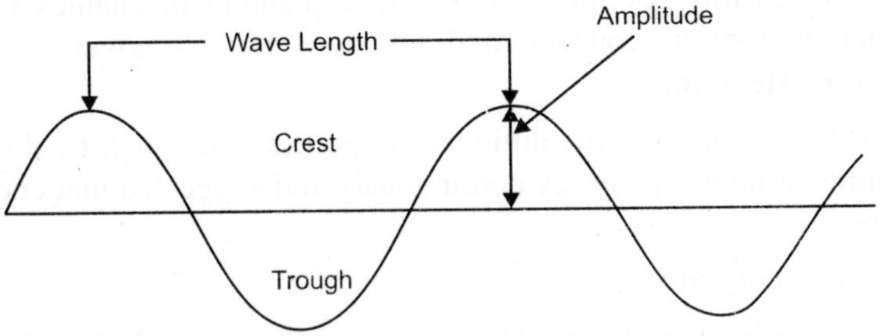


**Fig. 1.7** Wavelength and amplitude

**Crest** means the highest position to which the propagation medium rises while trough is the lowest position.

**Wave number.** It is defined as the total number of waves which, can pass through, a space of one cm. It is denoted by $\ddot{v}$ and is expressed in $cm^{-1}$. Wave number is equal to the reciprocal of wavelength (λ, expressed in cm) *i.e.,* $\ddot{v} = 1/\lambda$ in cm.

**Frequency.** It is defined as the number of waves or cycles which can pass through a point in one second. It is denoted by the letter $\ddot{v}$ (niu) and is expressed in cycles per second or in Hertz. The frequency of a radiation is inversely proportional to its wavelength or $\upsilon \propto 1/\lambda$ cm. Smaller the value of wavelength of a radiation, greater will be its frequency $\upsilon = c/\lambda$ where $c$ is the constant = velocity of light = $3 \times 10^{10}$ cm $sec^{-1}$.

**Amplitude.** It is the maximum height of the crest or depth of the trough. It is denoted by the Letter α.

**Velocity.** It is the distance covered by the waves in one second.

$$\text{velocity} = \text{frequency} \times \text{wavelength}$$

**Energy.** Energy of a wave of the particular radiation can also be calculated by applying the relation.

$$E = h\upsilon = h \cdot \frac{C}{\lambda}$$

The energy of light radiation can be calculated in ergs which can also be converted in kcal $mole^{-1}$ or in kJ $mole^{-1}$. The basic relationships of energy in calories per mole to frequency and wavelength are given by the expressions $E = Nh\nu = Nh\ c/\lambda$ where N is the Avogadro's number and E is the energy absorbed in ergs. The energy in electron volts is given by $ev = 1/8.066\lambda$ where λ is the wavelength measured in cm; one electron volt = 23.06 k cal/mole.

### 1.4.4 Elementary Quantum Theory of Max Planck

One of the biggest surprises of 20th century physics was the discovery that classical mechanics (the mechanics of macroscopic particles) is an approximation: it is inapplicable to like size of atoms and has to be replaced by Quantum mechanics. Until the present century it was assumed that the classical mechanics was applied to objects as small as atoms. Experimental evidence was accumulated, however, which showed that classical mechanics failed when it was applied to very small particles. Classical physics was thought to be wrong in allowing systems to posses arbitrary amounts of energy. When this key idea was pursued quantum mechanics was discovered and it was in 1926 when appropriate concepts and equations were discovered to describe the new mechanics: **Quantum Mechanics**.

Max Planck (1901) proposed a revolutionary hypothesis in which he discarded the precept that an oscillator emits or takes up energy continuously and suggested that energy changes occur in discrete amounts.

The postulates of this theory are:

- The energy is emitted or absorbed by a body not continuously but discontinuously in the form of small packets or stated otherwise an oscillator has definite energy levels $\varepsilon_0$, $\varepsilon_1$, $\varepsilon_2$, $\varepsilon_3$,....., $\varepsilon_i$ etc.
- Each packet of energy is called a quantum. A quantum of energy emitted in the form of light is known as photon.
- The energy of photon is not fixed. It is directly proportional to the frequency of light $\varepsilon \propto \nu$ or $\varepsilon = h\nu$ where $h$ is the Planck's constant, having the dimensions of energy × time (a quantity called 'action') = $6.625 \times 10^{-27}$ erg second (in C.G.S. unit) or else it can be stated that the oscillator emitting a frequency $\nu$ can only radiate in units or quanta of the magnitude $h\nu$, where $h$ is a fundamental constant of nature.

$$\varepsilon = h\nu$$

- This really amounts to introduction of the concept of atomicity in the realm of energy.
- A body can emit or absorb a photon of energy or some integral multiples of it *i.e.*, energy levels of the oscillator can only be integral multiples of a quantum.

  *i.e.*, $E_n = n\varepsilon = nh\nu$ where $n$ is an integer.

### 1.4.5 Postulate's of Bohr's Theory

- Each orbit around the nucleus is associated with a definite amount of energy and the orbits are therefore called energy levels or main energy shells. These shells are numbered as 1, 2, 3,..... starting from the nucleus and are designated by capital letters : K, L, M, ... respectively. The energy associated with a certain energy level increases with increase of its distance from the nucleus. Thus if $E_1$, $E_2$, $E_3$, .... denote the energies associated with the energy levels numbered as 1(K-shell), 2 (L-shell), 3 (M-shell)..., these are in order $E_1 < E_2 < E_3 <$ ..... Thus an outer energy level has higher energy than inner energy level. While revolving around the nucleus in a fixed orbit, the electron neither losses (*i.e.*, emits) nor gains (absorbs) energy *i.e.*, its energy remains constant as it is revolving in a particular orbit. Under this condition the atom as a whole is said to be in a state of stationary energy state or simply in a stationary state.

  Energy is however emitted or absorbed by an atom, when an electron jumps from one energy level to the other. The amount of energy ($\Delta E$) emitted or absorbed in this type of jump (transition) is given by Planck's equation.

Thus, $\Delta E = h\upsilon$

where $\upsilon$ = the frequency of the energy (radiation) emitted or absorbed.

- Although there are infinite number of circular concentric orbits in which an electron may be expected to move about the nucleus, the electron can move only in that orbit in which the angular momentum of the electron is quantised *i.e.,* the angular momentum of the electron is a whole number multiple of $h/2\pi$. This is kncwn as principal of quantisation of angular momentum according to which $mvr = nh/2\pi$, where m is the mass of the electron, $v$ is tangential velocity of the electron in its orbit, $r$ is the distance between the electron and nucleus and $n$ is a whole number which has been called the principle quantum number by Bohr. It is the number of the orbit in which the electron is revolving and can have the values 1, 2, 3,.... for the main energy levels numbered as 1(K-shell), 2 (L-shell), .... starting from the nucleus.

### 1.4.6 General Features of Spectroscopy

The origin of the spectral lines in molecular spectroscopy is the emission or absorption of a photon when the energy of the molecule changes. The difference from atomic spectroscopy is that a molecule's energy can change not only as a result of electronic transition but also its rotational and vibrational states may change. This means that the molecular spectra are more complex that atomic spectra; but also contain information relating to more properties such as bond strength and molecular geometry. The field of spectroscopy is divided into emission and absorption spectroscopy. An emission spectrum is obtained by spectroscopic analysis of some light source such as flame or an electric arc. This phenomena is primarily caused by the excitation of atoms by thermal or electrical means; absorbed energy causes electrons in the ground state to be promoted to a state of higher energy. The life time of electrons in this meta stable state is short, and they return to some lower excited state or to the ground state; the absorbed energy is released as light. The transmission form higher to a lower energy state and subsequent emission of excess energy as photon of frequency $\upsilon$ is given by $E_1 - E_2 = h\,\upsilon$. This relation is often expressed in terms of $c = \upsilon\lambda$ or the wave number $\ddot{v} = \upsilon/c$. (The relations of frequency, wavelength and wave number has already been discussed previously). However, in some cases the excited state sometimes may have appreciable life times such that emission of light continues after the excitation has ceased; such a phenomenon is called 'phosphorescence'.

When the radiation emitted by the excited substance are analysed by spectrograph(prism), a discontinuous spectra consisting of a series of sharp lines with dark lines in between result and is called line spectrum. In absorption spectroscopy the absorption of incident radiation is monitored as it is swept over a range of frequencies, the presence of an absorption at a frequency $\upsilon$ signifying the presence of two energy levels separated by $h\upsilon$ as expressed by $E_1 \sim E_2 = h\upsilon$. An absorption spectrum is obtained by placing the substance between the spectrometer and some source of energy that provides electromagnetic radiation in the frequency range being studied. The spectrometer analyses the transmitted energy relative to the incident energy for a given frequency. Again the high energy states are usually short lived. The major fate of absorbed energy in the ultra violet region is re-emission of light. Occasionally the absorbed energy may cause photo chemically induced reactions. Although the mechanism of energy absorption is different in the UV, IR and nuclear magnetic resonance (NMR) regions the fundamental process is the absorption of certain amount of energy. For a given excitation process, a molecule absorbs only one discrete amount of energy, and hence absorbs radiation of only one frequency. If this were the case with all molecules of a substances, one would observe a series of absorption lines. However, a group of molecules exists in a number of different vibrational and rotational states; each state differing from another by a relatively small amount of energy. Thus a grouping of molecules absorbs energy over a small range and gives rise to an absorption band or peak.

Emission and absorption spectroscopy give the same information about energy level separations but practical considerations generally determine which technique is employed. Absorption of ultra violet and visible light is chiefly caused by electronic excitation, the spectrum provides limited information about the structure of the molecule. In order to obtain useful information from UV and visible range spectrum of a compound the wavelength of maximum absorption ($\lambda_{max}$) and the intensity of absorption must be measured accurately. The mechanics of measurement is thoroughly dealt within Article 1.3.

### 1.4.7 General Discussion and Elementary Theory of Flame Spectrometry (Atomic Absorption Spectrometry and Flame Photometry)

If a solution containing a metallic salt (or some other metallic compound) is aspirated into a flame (acetylene burning in air), a vapour which contains atoms of the metal may be formed. Some of these gaseous metal atoms may be raised to an energy level which is sufficiently high to permit the emission of radiation characteristic of that metal e.g., the characteristic yellow colour imparted to the flames by compounds of sodium. This is the basis of flame emission spectroscopy (FES), often referred to as flame photometry. However, a much larger number of the gaseous metal, atoms will normally remain in an unexcited state, or in other words, in the ground state. These ground state atoms are capable of absorbing radiant energy of their own specific resonance wavelength, which in general is the wavelength of the radiation that the atoms would emit if excited from the ground state. Hence if light of the resonance wavelength is passed, through a flame containing the atoms in question, then part of the light will be absorbed and the extent of absorption will be proportional to the number of ground state atoms present in the flame. This is the underlying principle of atomic absorption spectroscopy (AAS).

Let us consider the simplified energy level diagram shown in Fig. 1.8 where $E_0$ represents the ground state in which the electrons of a given atom are at their lowest energy level and $E_1$, $E_2$, $E_3$ etc. represent higher or excited energy levels. Transition between two quantised energy levels, say from $E_0 \rightarrow E_1$ corresponds to absorption of radiant energy, and the amount of energy absorbed ($\Delta$ E) is given by Bohr's equation

$$\Delta E = E_1 - E_0 = h\nu = h\frac{c}{\lambda}$$

where;

$c$ = velocity of light

$h$ = Planck's constant

$\nu$ = frequency

$\lambda$ = wavelength of radiation absorbed.

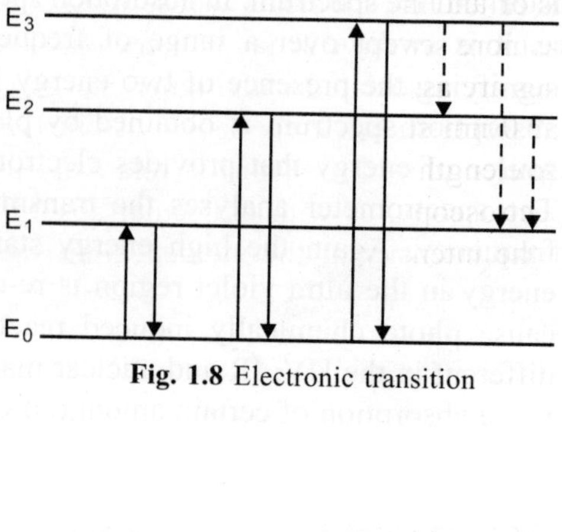


**Fig. 1.8** Electronic transition

Clearly the transition from $E_1 \rightarrow E_0$ correspond to the emission of radiation of frequency $\upsilon$. Since an atom of a given element gives rise to a definite, characteristic line spectrum, it follows that there are different excitation states associated with different element. The consequent emission spectra involve not only transitions from excited states to the ground, state e.g., $E_3 \rightarrow E_0$, $E_2 \rightarrow E_0$ (as indicated by bold lines in Fig. 1.8), but also transitions such as $E_3 \rightarrow E_2$, $E_3 \rightarrow E_1$ (as indicated by the dotted lines). Thus it follows that emission spectrum of a given element is quite complex. Theoretically it is always possible for absorption of radiation by already excited states to occur; e.g., $E_1 \rightarrow E_2$, $E_2 \rightarrow E_3$ etc. But in practice the ratio of excited to ground state atoms is extremely small, and thus the absorption spectrum of a given element is usually only associated with transitions from the ground state to higher energy states and is thus much simpler in characteristics than the emission spectrum. The relationship between ground state and excited state population is given by the Boltzmann equation.

$$\frac{N_1}{N_0} = \left(\frac{g_i}{g_0}\right) e^{\frac{-\Delta E}{kt}}$$

where

$N_1$ = number of atoms in the excited state,

$N_0$ = number of atoms in the ground state,

$g_i/g_0$ = ratio of statistical weights for excited and ground states,

$\Delta E$ = energy of excitation = $h\upsilon$,

$k$ = the Boltzmann's constant,

T = absolute temeperature (K).

It can be seen, from the equation that the ratio $(N_1/N_0)$ is dependent upon both the excitation energy $\Delta E$ and the temperature T. An increase in temperature and a decrease in $\Delta E$ (*i.e.,* when dealing with transitions which occur at longer wavelengths) will both result in a higher value for the ratio $N_1/N_0$.

Atomic absorption spectroscopy is less prone to inter element, interferences than is flame emission spectroscopy. Further due to high proportion of ground state to excited state atoms it would appear that atomic absorption spectroscopy should also be more sensitive than flame emission spectroscopy. However, in this respect, the wavelength of the resonance line is a critical factor and the elements whose resonance lines are associated with relatively low energy values are more sensitive as far as flame emission spectroscopy is concerned than those whose resonance lines are associated with higher energy values. Thus sodium with an emission line of wavelength 589.0 nm shows great sensitivity in flame emission spectroscopy, whereas zinc (emission line wavelength = 213.9 nm) is relatively insensitive. It should be noted that in atomic absorption spectroscopy, as with molecular absorption, the absorbance A is given by the logarithmic ratio of the intensity of the incident light signal $I_0$ to that of the transmitted light $I_t$ *i.e.,*

$$A = \log \frac{I_0}{I_t} = KLN_0$$

where

$N_0$ = concentration of the atoms in the flame (number of atoms per $cm^3$),

L = path length, through the flame (cm),

K = constant related to the absorption coefficient.

With flame emission spectroscopy, the detector response E is given by the expression

$$E = K\alpha C$$

where

$k$ = related to a variety of factors including the efficiency of atomisation and of self absorption,

$\alpha$ = the efficiency of atomic excitation,

C = the concentration of the test solution.

### 1.4.8 Flame Photometry

When a substance is heated, it emits radiant energy. The emission becomes stronger with greater excitation of the molecules/atoms. This energy (electromagnetic radiation) composed of radiation is the emission spectrum of the substance. There are three kinds of emission spectra:

- **Continuous spectrum**, given out by incandescent solids, consisting of continuous wavelength range, where individual, lines are absent.
- **Band spectrum** emitted by excited molecules/atoms consisting of individual bands which are actually composed of groups of lines very close to one another.
- **Line spectrum** originating from excited atoms or atomic ions (excluding poly atomic ions or radicals). These spectra consists of distinct and often widely spaced lines.

A flame photometer is an instrument in which the intensity of the filtered radiation from the flame is measured with a photoelectric detector. The filter interposed between the flame and the detector, transmits only a strong line of the element.

Analytical flame photometry is based on the measurement of the intensity of the characteristic line emission of the element to be determined (Jackson 1973). When a solution of a salt is sprayed into a flame (acetylene, propane or liquefied petroleum gas) the salt gets separated into its component atoms because of the high temperature. The energy provided by the flame excites the atoms to higher energy levels. Actually the orbital electrons are shifted to higher planes from their normal orientation. When the electrons return back to ground state or unexcited state, they emit their characteristic radiation. Since the excitation can be to different levels, light (electromagnetic radiation) of several wavelengths can be emitted. However, the intensity of the wavelength corresponding to the most probable transition will be the highest. For each element such characteristic lines have already being well identified. Each individual, atom emits one quantum of radiation, therefore, the intensity of radiation emitting from the flame will be proportional to the number of atoms in the flame, that is, to the concentration of the particular element in the flame. This concentration is in turn directly related to the content of the element in the test solution.

The instrumental set up for flame photometric analysis consists of three parts.

- **Nebulizer burner system** which converts the test solution to gaseous atoms. The function of nebulizer is to produce a mist or aerosol of the test solution.
- **Monochromation system** (filter, prism) that separates out the analytical wavelength, from other radiations; and
- **Photometric system** for measuring the intensity of the emitted radiation.

### Experimental

A series of standard solutions are prepared and the intensity of emission determined for each concentration after zero setting of blank and hundred setting of the maximum concentration. The intensity of emissions from the test solutions is measured simultaneously and the concentration of the element is read from the calibration curve.

In a single beam instrument referred to as direct reading type, comprises only one set of optics Light emitted from the core of the flame just above the inner cone ions is collected by a reflector and focussed by a lens of heat resistant glass through interchangeable optical filters on to a single photo detector. Alternatively, light from the burner passes into the monochromator and radiation leaving the exit slit is focussed on to the photo detector unit, (Jackson 1973).

Flame photometers are intended, primarily for the analysis of sodium and potassium and also for calcium and lithium *i.e.,* elements which have an easily excited flame spectrum of sufficient intensity for detection by a photocell. In actual practice, air at a given pressure is passed into an atomiser and the suction this produces draws a solution of the sample into the atomiser, where it joins the air stream as a fine mist, and passes into the burner. At this stage the air meets the fuel gas supplied to the burner at a given pressure and the mixture is burnt. Radiation from the resulting flame passes through a lens then through an iris diaphragm and finally through an optical filter which permits only the radiation characteristics of the element under investigation to pass through to the photocell. The output from the photocell is measured, on a suitable galvanometer. The flame is protected by a chimney to protect, it from draughts. The optical path, from the chimney to the photocell is enclosed in a light tight box. Commonly used, flame photometers models are EEL-Cornings, Coleman, Systronics, ELICO etc. (direct reading absorption filters, barrier layer cell) and Baird model KY-2, Lange model 4 etc. (internal standard, interference filters, barrier layer cell).

### 1.4.9 Care and Maintenance

Flame photometer give trouble free service for years if handled properly in spite of being a very sensitive instrument. The gas and air pressure must remain steady during the operation of the instrument and the connections and regulators should be checked before use. It is very important that only very clear solutions/extracts are fed in to avoid any chocking of the capillary. Burners should be cleaned periodically. While closing down, the gas supply should be first turned off before closing compressor. When large salt concentrations or strong acid solutions are fed (which should be avoided as far as possible) distilled water should be run for some time and the burner cleaned immediately after use. The galvanometer assembly must not be touched and any repairs must be done at good service centres.

*Schematic representation of instrumentation for flame spectro photometric procedures*:

The components included within the frame drawn in dotted lines represents the apparatus required for flame emissions spectroscopy. For atomic absorption spectroscopy there is an additional requirement of a resonance line source.

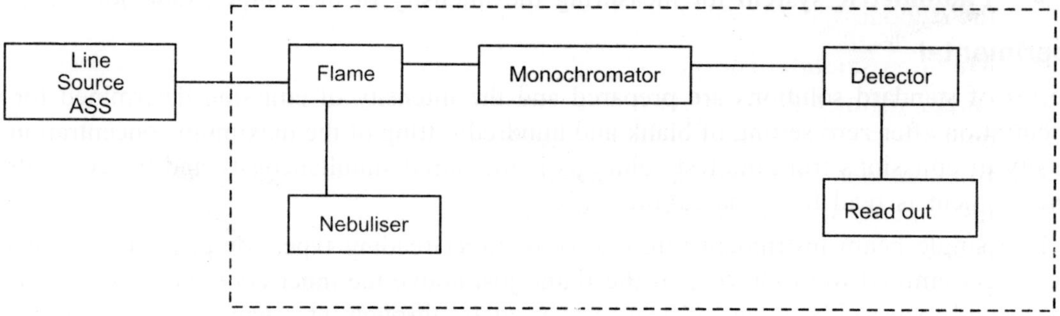


**Fig. 1.9** Schematic representation of flame spectrometric procedure

For flame spectroscopy an essential requirement is that the flame used, shall produce temperatures in excess of 2000 K. Flame temperatures with various fuels are shown below.

| Fuel gas | Temperature (K) | |
|---|---|---|
| | Air | Nitrous oxide |
| Acetylene | 2400 | 3200 |
| Hydrogen | 2300 | 2900 |
| Propane | 2200 | 3000 |

The concentrations of the gaseous atoms within the flame, both in ground and in excited states may be influenced by (*a*) the flame composition and by (*b*) the position considered within the flame. As far as flame composition is concerned, it may be noted that an acetylene-air mixture is suitable for determination of most of the metals, but propane-air flame is to be preferred for metals which are easily converted into an atomic vapour state. For metals such as aluminium and titanium, the higher temperature of the acetylene-nitrous oxide flame is essential. With regard to position within the flame, it can be shown that in certain cases the concentration of atoms may vary widely if the flame is moved either vertically or laterally relative to the light path from the resonance line source.

## 1.4.10 Atomic Absorption Spectrophotometer (Instrumentation and Experimental)

- **Nebulizer-burner system.** The purpose of the system is to produce uniformly fine fog of droplets from the test solution. The burner has a long and narrow slot at the top so that the flame provides a long absorption path for the incident radiation. The fuel-oxidant system used may be acetylene air, acetylene-nitrous oxide, hydrogen-air etc.

- **Resonance line sources.** For atomic absorption spectroscopy a resonance line source is required which is the hollow cathode lamp. Far any given determination the hollow cathode lamp used has an emitting cathode of the same element as that being studied in the flame. The cathode is in the form of a cylinder and the electrodes are enclosed in a borosilicate or quartz envelope which contain an inert gas (neon or argon) at a pressure of approximately 5 torr. The application of high potential across the electrodes causes a discharge which creates ions of the noble gas. These ions are accelerated to the cathode and on collision, excite the cathode element to emission.

- **Monochromator.** The general purpose of the monochromator is to select a given emission line and to isolate it from other lines and occasionally from molecular band emissions. In AAS the function of the monochromator is to isolate the resonance line from all non absorbed lines emitted by the radiation source. In most commercial instruments diffraction gratings are used because the dispersion provided by a grating is more uniform than that given by prisms and consequently grating instruments can maintain a higher resolution over a longer range of wavelengths.
- **Detectors and read out system.** In atomic absorption spectrophotometers in view of the improved, spectral sensitivity required photo multipliers are employed. The output from the detector is fed to a suitable read out system and in this connection is must be kept in mind, that the radiation received by the detector originates not only from the resonance line which has been selected, but may also arise from emission within the flame. The emission can be due to atomic emission arising from atoms of the element under investigation and may also arise from molecular band emissions. Hence, instead of an absorption signal intensity $I_A$ the detector may receive a signal of intensity ($I_A$ + S) where S is the intensity of emitted radiation. Since only the measurement arising from the resonance line is required, it is important that this be distinguished from the effects of flame emission. This is achieved by modulation of the emission from the resonance line source by either a mechanical chopper device or electronically by using an alternating current signal, appropriate to the particular frequency of the resonance line, and the detector amplifier is then tuned to this frequency. In this way, the signals, arising from the flame, which are essentially d.c. in character are effectively removed. The read out systems available include meters, chart, recorders and digital display.

## Experimental

The absorbance values corresponding to known standard solution are recorded after necessary blank settings and a calibration curve is prepared. The absorbance of the test solution is determined and concentration read off from the calibration curve.

A calibration curve for use in atomic absorption or in flame emission measurements is plotted by aspirating concentrations of the element to be determined, measuring the absorption (emission) of each solution, and then constructing a graph in which the measured absorption (emission) is plotted against the concentration of the solutions. For all absorbance measurements, the readings must be taken after the instrument zero has been adjusted against a blank for which double distilled water is used normally. If we are dealing with a test solution which contains a single component then the standard solutions are prepared by dissolving a weighed quantity of a salt of the element to be determined in a known volume of double distilled water in a volumetric flask. If necessary the test solution must be suitably diluted and dilution factors are to be noted for final calculation.

### Types of AAS

Some of the established manufacturers of different types and models of atomic absorption spectrophotometer are: Perkin-Elmer, Cooperation, Norwalk, Conn. USA; Hitachi Instruments Co., Tokyo, Japan; Atomic Absorption and Electronics Corporation, New York; Bausch & Laumb Inc., Rochester New York; Hewlett-Packard, Dayton, Ohio and many others.

### Salient points regarding operation and maint nance of AAS

Since AAS is one of the most sophisticated, high precision and expensive type of measuring instrument, the AAS requires very careful handling and maintenance.

The light source is very important and critical component. There should not be any erratic fluctuations in intensity. Any fluctuations in the line voltage must be taken care of. While extinguishing the flame the fuel supply should be first turned off before closing the air supply. The fuel and air pressure must be adjusted to the value recommended in the instruction manual, furnished by the manufacturer. The slit and wavelength must be set properly for each determination. Various aspects to be considered for optimum working conditions are: steady rate of atomization, the intensity of emission, the resolution of spectral lines (or bands) from each other, the magnitude of background radiation, and steadiness of readings. The AAS gives reliable and trouble free service only with careful handling. Erratic readings originates from fluctuations in voltage, gas and air pressure and also the hallow cathode lamp becoming weak and due to prolonged use. For repairing purpose, the instrument should never be opened and must be carried out by authorised agents of the manufacturers.

## 1.4.11 Interferences

Several factors may affect the flame emission of a given, element and lead to interference with the determination of the concentration of given element. The factors may be broadly classified as **(a) spectral interferences and (b) chemical interferences.**

### *Spectral interferences*

Spectral interferences in AAS arise mainly from overlap between the frequencies of a selected resonance line with lines emitted by some other element; which arises because in practice a chosen line has in fact a finite 'band-width'. With flame emission spectroscopy, there is greater likelihood of spectral interferences where line emission of the element to be determined and those due to interfering substances are of similar wavelength, than with atomic absorption spectroscopy. Some of such interferences are eliminated by improved resolution of the instrument e.g., use of prism rather than a filter, but in certain cases it may be necessary to select other, non interfering lines for the determination. Additionally some interference may arise from the emission band spectra produced by molecules or molecular fragments present in the flame gases: in particular, band spectra due to hydroxyl and cyanogen radicals arise in many flames.

### *Chemical interferences*

The production of ground state gaseous atoms which is the basis of flame spectroscopy may be inhibited by two main forms of chemical interference:

- By stable compound formation
- By ionisation.

- **Stable compound formation** leads to incomplete dissociation of the substance to be analysed when placed in flame for e.g., determination of calcium in presence of sulphate and phosphate. Chemical interferences can usually be overcome in one of the following ways:

  *Increase in flame temperature* often leads to the formation of free gaseous atoms for e.g., aluminium oxide is more readily dissociated in acetylene nitrous oxide flame than it is in air acetylene flame.

  *By use of releasing agents* Considering the reaction M – X + R = R – X + M, it becomes evident that an excess of the releasing agent (R) will lead to an enhanced concentration of the required gaseous metal atoms (M) which will be of special significance if the product R – X is a stable compound. Hence in the determination of calcium in presence of phosphate the addition of excess of strontium chloride to the test solution will lead to the formation of strontium phosphate and the calcium can then be determined in an acetylene – air flame without any interference due to phosphate. Also addition of EDTA to a calcium solution before analysis may increase the sensitivity of the subsequent flame spectrophotometric determination which may be due to the formation of an EDTA complex of calcium which is readily dissociated in the flame.

  *Extraction of the analyte or of the interfering element(s)* is an obvious method of overcoming the effect of 'interferences'. It is sufficient to perform a simple solvent extraction to remove the major portion of an interfering substances so that at the concentration at which it then exists in the solution, the interference becomes negligible.

- **Ionisation of the ground state gaseous atoms within a flame**, $M = M^+ + e^-$ will reduce the intensity of emission of the atomic spectral lines in a flame emission spectroscopy or will reduce the extent of absorption in atomic absorption spectroscopy. It is therefore, very necessary to reduce the possibility of ionisation occurring to a minimum and as an obvious precaution is to use a flame operating at the lowest possible temperature which is satisfactory for the element to be determined. For instance, the high temperature of acetylene nitrous oxide flame may result in appreciable ionisation of elements such as the alkali metals and of calcium, strontium etc. The ionisation of the element to be determined may also be reduced by the addition of an excess of an ionisation suppresant, which is essentially a solution containing a cation having a lower ionisation potential than that of the analyte. Thus, for example a solution containing potassium ions (2000 ppm say) if added to a solution containing calcium, barium or strontium creates an excess of electrons when the resulting solution is nebulised into the flame and this has the result that the ionisation of the metal to be determined is virtually completely suppressed.

In addition to the compound formation and ionisation effects which have been already discussed, it is also necessary to take account of so called matrix effect. These are predominantly physical factors which will influence the amount of sample reaching the flame and are related to factors such as viscosity, the density, the surface tension and the volatility of the solvent used to prepare the test solution.

In some circumstances interference may result from molecular absorption. Thus, for example, in an acetylene-air flame a high concentration of sodium chloride will absorb radiation at wavelengths in the neighbourhood of 213.9 nm which is the wavelength of the zinc resonance line. Hence, sodium chloride would represent an interference in the determination of zinc under these conditions. Such interferences can usually be avoided by choosing a different resonance wavelength for carrying out the determination or else by using a different flame so that the operating temperature is increased thus leading to dissociation of the interfering molecules. It may also be noted that interference referred to as background absorption, which arises from the presence in the flame of gaseous molecules, molecular fragments is dealt with in many modern instruments by the incorporation of a background correction facility. Usually a background corrector is incorporated which takes the form of high intensity deuterium arc lamp producing an emission continum which travels the same double beam path, as does the light from the resonance source. The background absorption affects both the sample and reference beams and hence when the ratio of the intensities of the two beams is taken, the background effect is eliminated.

With regard to the relative merits of FAAS and FES procedures, it may be stated in general that FAAS is more selective technique than FES, and in terms of sensitivity it is also to be preferred when we are dealing with lines of wavelengths less than 350 nm. However, for lines of wavelengths appreciably greater than 350 nm the FES is more sensitive technique.

### 1.4.12 Safety Practices

Before performing any experimental work with either a flame (emission) photometer or an atomic absorption spectrophotometer the following guidelines on safety practices must be studied. These recommendations are a summary of the code of practice recommended by the Scientific Apparatus Makers Association (SAMA) of USA.

- Ensure that the laboratory in which the apparatus is installed is well ventilated and is provided with an adequate exhaust system having air tight joints on the discharge side.
- Gas cylinder must be fastened securely in an adequately ventilated room well away from any heat or ignition sources. The cylinder must be clearly marked so that the contents can be easily identified.
- When the equipment is turned off, close the fuel gas cylinder valve tightly and bleed the gas line to the atmosphere via the exhaust system.
- The piping which carries the gases from the cylinders must be securely fixed in such a position that it is unlikely to suffer damage.
- The following special precautions should be observed with acetylene:
  - Never run acetylene at a pressure higher than 15 p.s.i; at higher pressures acetylene can explode spontaneously.
  - Avoid contact between gaseous acetylene and silver mercury or chlorine.
  - Avoid use of copper tubing. Use tubing made from brass containing less than 65%

Copper, from galvanized iron or from any other material that does not react with acetylene.

- ➢ Never run acetylene cylinder after the pressure has dropped to 50 p.s.i., at lower pressures the gas will be contaminated with acetone.

- A nitrous oxide cylinder should not be used after the regulator gauze has dropped to a reading of 100 p.s.i.
- Never view the flame or hollow cathode lamp directly. Protective eye wear should always be worn. Safety spectacles will usually provide protection from ultraviolet light and will also provide protection for the eyes in the event of apparatus being shattered by explosion.
- Care must be exercised when using volatile inflammable organic solvents for aspiration into the flame
- A burner which utilises a mixture of fuel and oxidant gases and which is attached to a waste vessel (liquid trap) should be provided with a U-shaped connection between the trap and the burner chamber. The head of liquid in the connecting tube should be greater than the operating pressure of the burner, if this is not achieved, mixtures of fuel and oxidant gas may be vented to the atmosphere and forms an explosive mixture. The trap should be made of a material that will not shatter in the event of an explosive flash back in the burner chamber.
- Never leave a flame unattended.

## 1.5 INDUCTIVELY COUPLED PLASMA-ATOMIC EMISSION SPECTROSCOPY (ICP)

A new analytical technique called Inductively Coupled Plasma Atomic Emission Spectroscopy (ICP-AES) is used for simultaneous multi-element analysis of biological material and soils. Elements such as Al, P,S and B which are either poorly measured at low concentrations or not measured with AAS, are readily determined with higher sensitivity by ICP-AES. Moreover, ICP-AES has multi-element capability at all concentration levels.

### Principle

ICP-AES is based on the observation of atomic emission spectra when samples in the form of an aerosol, thermally generated vapour or powder are injected into an inductively coupled plasma atomization and excitation source. By definition, plasma refers to a hot gas in which a significant fraction of their atoms or molecules are ionized. Plasmas are electrically conducted and have been referred to as electrical flames, as no combustion takes place. This is because instruments using a plasma source generally used inert argon gas.

The ICP is produced by passing initially ionized argon gas through a quartz torch located within an induction coil (Cu coil) which is connected to a radio frequency (RF) generator. The RF generator produces 1.5 to 3 kW power at a frequency of 27.1 MHz. An oscillating magnetic field is formed within the quartz torch in response to the RF energy passing through the coil. Electrons and ions passing through the oscillating electromagnetic field flow at high acceleration rates within the quartz torch space. As argon gas enters the magnetic field associated with the induction coil, its atoms collide with the accelerated ions and electrons resulting in the ionization of the argon gas. These collisions give rise to ohmic heating, which produces plasma with temperature ranging from 6000 to 10000 K. The resultant plasma is contained within the torch by means of argon flow.

The method of presenting the sample to the plasma is similar to that used in flame atomic absorption. The liquid sample is aspirated into the plasma through a nebulizer system by using argon carrier gas at a rate of about 1 L argon $min^{-1}$. The prevailing high temperature in the plasma leads to complete vaporization, atomization and excitation of the element to be analyzed. The exited neutral atoms or ions of the sample emit radiation of characteristic wavelengths with the intensity directly proportional to the element concentration in the sample. Intensity of the emitted radiation is measured by the spectrophotometer component of the ICP-AES.

## Preparation of samples for ICP-AES

Digestion procedure of soil and plant samples for total elemental analysis by ICP-AES is similar to that used for various emission instruments. Universal/multi-element soil extractants can also be used for the extraction of soil samples (Soltanpour *et al.,* 1998).

## Preparation of standard solution

Stock standard solution containing 1000 mg $l^{-1}$ of an element may be prepared as described for AAS. However, multi-element working standard solutions (secondary standards) should be made in such a way that these contain maximum number of the elements compatible with stability considerations and match with the sample solutions in kind and strength of acids. In soil and plant analysis, one set of secondary standards is required for each multi-element extracting solution, and one for each soil and plant digest. McQuaker *et al*., (1979) devised a calibration scheme for 30 elements that satisfies the needs of analysts engaged in soil, water, tissue, and particulate matter analysis.

## Operation of ICP-AES

The general principles for the operation of ICP-AES are similar to that for AAS. Microprocessors are used for the automatic control of instrument components, set up and optimization of required operational parameters, instrument calibration, and manipulation and storage of data through key commands. Since the actual operation of different ICP spectrometers varies with the type, make and computer software of the instrument, the readers are advised to consult operation manual provided by manufacturer of the instrument. The manufacturers/dealers may also provide required operational training to the user.

## Precautions

- Filter soil extracts/plant digests with Whatman No. 42 filter paper to prevent clogging of the nebulizer.
- Use high salt nebulizer (Babington type) or standards and samples having very low salt content to prevent clogging of the nebulizer tip.
- Avoid mixing of chemicals that cause precipitation during the preparation of multi-element working standard solutions.

•••

# 2 Soil Physics

## 2.1 PARTICLE SIZE DISTRIBUTION

### General discussion and principle

Particle size analysis separates the inorganic mineral portion of the soil into classified grades according to particle size and determines their relative proportion by weight. The determination involve three distinct stages viz.

- Removal of cementing agents like organic matter and calcium ions and complete dispersion of the soil samples in an alkaline medium. Dispersion is brought about effectively by a combination of physical and chemical means. Organic matter, free iron oxide and flocculating ions such as calcium and magnesium are combated chemically and the dispersion process is speeded up by addition of dispersing agents and mechanical shaking.
- Separation of the coarse sand fraction (International system) by wet sieving followed by separation of total sand fraction (USDA) system by dry sieving.
- Determination of the clay fraction (both systems) and silt fraction (International system) in the dispersed sample by pipetting known volume of sample or by measuring specific gravity with a special hydrometer, each after calculated times which ensure that the required fraction is being determined.

The velocity of sedimentation is related to the particle size by Stoke's law. The law dictates the rate of fall of small spherical particles through a viscous fluid. (Note: The soil particles are dispersed in water after removal of cementing agents viz. organic matter, calcium carbonate etc. by using dispersing agent like sodium hexametaphosphate or calgon). These actually increases the zeta potential ($\xi$) of the medium. Particles greater than 0.05 mm are separated by sieving. Using Stoke's law the depth to which particles larger than 0.02 mm in equivalent diameter will settle in 5 minutes and the depth to which particles larger than 0.002 mm will settle in 5 hrs. 30 minutes are calculated. Aliquots pipetted from these depths after the appropriate times will contain only particles smaller than the two cut off diameters mentioned above.

### Stokes' law

A particle falling in a vacuum will encounter no resistance, as it is accelerated by gravity and hence its velocity increases as it falls. A particle falling in a fluid on the other hand will encounter a frictional resistance proportional to the product of its radius and velocity and to the viscosity of the fluid.

The resisting force or the viscous drag due to friction was shown by Stokes G.G. (1851) to be

$$F = 6\pi\eta r v \qquad \text{...(2.1.1)}$$

where

$\eta$ = viscosity of the fluid,

$r$ = radius of the particle,

$v$ = velocity of the particle.

Initially, as the particle begins to fall, its velocity increases. Eventually a point is reached at which the increasing resistance force equals the constant downward force, and the particle then continues to fall without acceleration at a constant velocity, known as terminal velocity $v_t$.

The downward force due to gravity

$$F = mg = 4/3\ \pi r^3\ (\rho - \sigma)g \qquad \text{...(2.1.2)}$$

where $\rho$ = density of the spherical material,

$\sigma$ = density of liquid.

Setting the two forces equal, we get Stoke's law

$$4/3\ \pi r^3\ (\rho - \sigma)g = 6\pi\eta r v \qquad \text{...(2.1.3)}$$

or $$v_t = \{(2/9\eta\ (\rho - \sigma)g)\ r^2\} \qquad \text{...(2.1.4)}$$

or $$v_t = \mathrm{K}r^2 \qquad \text{...(2.1.5)}$$

where K = proportionality constant = $\{2/9\eta\ (\rho - \sigma)g\}$

Also $$v_t = \frac{\text{distance}\,(h)}{\text{time}\,(t)} \qquad \text{...(2.1.6)}$$

Therefore, $$h/t = \mathrm{K}r^2 = \{2/9\eta\ (\rho - \sigma)gr^2\} \qquad \text{...(2.1.7)}$$

Simplifying we get, $$t = \{9\eta - h)/2(\rho - \sigma)gr^2\} \qquad \text{...(2.1.8)}$$

as $r$ = diameter $(d)/2$,

equation (2.1.8) becomes

$$t = \{18\eta h)/(\rho - \sigma)gd^2\} \qquad \text{...(2.1.9)}$$

**Assumptions in Stokes' law**

- The particles must be large in comparison to liquid molecules so that Brownian movement will not affect the fall.
- The volume of liquid should be greater in comparison with the size of the particles. The fall of the particle must not be affected by proximity of the wall of the vessel or of adjacent particles.

- Particles must be smooth and rigid, a condition which is difficult to be fulfilled by soil particles.
- It is a well known fact that the soil particles are not spherical exactly but are irregularly shaped with a large number of plate-shaped particles present in the clay fractions. Since particles with different shapes fall with different velocities, the term equivalent or effective radius is used to overcome this difficulty in Stoke's law. Effective radius is defined as the radius of a sphere of the same materials which would fall with the same velocity as the particle in question.
- There must be no slipping between various particle – a postulate which is well fulfilled in case of soil particles due to the presence of water hull around them.
- The velocity of fall must not exceed a certain critical value so that, the viscosity of the liquid remains the only resistance to the fall.
- In addition to the effect of size and shape of the particles upon the applicability of Stoke's law in particle size analysis there are certain experimental limitations that must be considered in the use of this principle. Since the rate of fall varies inversely with the viscosity of the medium, it is important to maintain a known constant temperature during the analysis. A constant temperature also helps to prevent convection currents which might arise as a result of difference in temperature near the walls of the vessel and within the suspension. Such currents acts as a hindrance to uniform settling of the particles. In addition convection currents may also be set up during stirring which is more difficult to eliminate than those arising out of temperature variation.
- The density of the soil particle is another factor which affects the accuracy of Stoke's law. Density depend upon the mineralogical and chemical constitution of the particles as well as upon their degrees of hydration. Usually $\rho$ is taken to be 2.65 and $\sigma$, 1.00 for mechanical analysis.

### 2.1.1 International Pipette Method

#### Reagents

- Hydrogen peroxide ($H_2O_2$) 30% (100 vols)
- Dispersing agent: Dissolve 36 g sodium hexametaphosphate and 8 g sodium carbonate in one litre water
- Hydrochloric acid 2(N)

#### Apparatus

- Tall form beakers (600 mL) and watch glasses
- 400 mL beakers and watch glasses
- Buchner funnels
- Suction assembly
- Wide mouth reagent bottles (500–600 mL) with rubber stoppers.
- 1000 mL cylinders
- Mechanical shakers

- Stirrer or plunger for mixing, consisting of a circular brass disc (55 cm diameter) fastened to a 600 mm length of brass rod. The disc is pierced with 8–10 holes of 4–5 mm diameter.
- Thermometer
- Wash bottles
- Water bath
- Drying oven
- Desiccator
- Analytical balance
- Stop watch
- Weighing dishes (Aluminium – 40 mL)
- Robinson pipette stand, with rack and pinion arrangement for raising and lowering pipette and scale.
- Robinson pipette (20 mL)
- Sieves of diameter 4–5 inches (1.0 mm, 0.5 mm, 0.25 mm, 0.2 mm, 0.105 mm, 0.053 mm)

**Procedure**

- Weigh 10 g air dry soil in 600 mL beaker and add 25 mL water and swirl.
- Add 5 mL of hydrogen peroxide, cover the beaker for overnight.
- Next day keep it on a hot plate at about 70°–80°C adding 5ml portions of hydrogen peroxide at 1 hour interval with occasional stirring.
- Continue till large bubbles cease to evolve, to ensure complete oxidation of organic matter.
- Boil gently for about an hour to decompose excess peroxide.
- Allow the contents to cool and add 25 mL of 2(N) HCl (if soil contains more than 2% carbonate more HCl may be added).
- After the reaction is over filter the suspension through Whatman No. 40 filter paper using suction.
- Transfer all the soil to the filter paper. Wash the soil several times (hot water may be used) to free from acid and soluble material.
- Transfer the soil on the filter paper carefully to a 400 mL previously weighed dry and clean beaker using minimum amount of water.
- Evaporate the suspension to dryness on a hot plate or sand bath.
- Keep the beaker overnight in a hot-air over at 105°C. Cool the bearer in a desiccator and weigh quickly to the nearest mg.

  [*Note:* This weight minus weight of the beaker gives the weight of sample (X) in the nearest mg, *i.e.,* the base weight.]
- Now slake the soil with water and transfer it quantitatively to a 500–600 mL, reagent bottle.
- Add 10 mL of dispersing agent (sodium hexametaphosphate) and dilute it to about 300 mL. Stopper tightly and shake in a mechanical shaker for 8 hours.

- Support the 53μ sieve in a funnel over a one litre measuring cylinder.
- Wash the soil, through the sieve until the washings are clear. Make up the volume up to the mark with distilled water.
- Transfer the portion on the sieve (*i.e.,* particles larger than 0.05 mm) which is the sand according the USDA system) to a previously weighed evaporating basin, dry and weigh to the nearest mg.
- Dilute the suspension in the cylinder 2000 mL, mix with plunger and record the temperature in °C.
- Stir the suspension with the plunger using vertical stroke for about 2–4 minutes holding the cylinder firmly during upward movement.
- Use strong upward strokes near bottom, but move the stirrer cautiously near the top of the suspension to avoid spilling.
- Avoid swirling. Start the stopwatch immediately after the plunger is removed from the suspension.
- After 4 minutes 50 seconds bring the Robinson pipette stand (with the pipette) to the cylinder. Close the stopcock of the pipette and lower it until the tip just touches the surface of the suspension.
- Note the reading on the scale and calculate the reading for the appropriate depth of sampling for the 0.02 mm cutoff at the temperature of the suspension (Table 2.1.1).
- After 4 minutes 55 seconds gently lower the pipette up to the required depth. After exactly 5 min open the stopcock and draw in the suspension rapidly and smoothly uninterruptedly until the pipette is filled to about 1 cm above the stopcock.
- Close the stopcock immediately and raise the pipette from the liquid.
- Drain out the excess liquid in the pipette through the side hole of the stopcock.
- Collect the aliquot in a previously weighed aluminium dish (or a weighed beaker 100 mL say) and evaporate to dryness.
- Keep in an oven at 105°C overnight, cool in a desiccator and weigh quickly.
- The weight (Y) of this fraction is the weight of clay plus international silt. Rinse the pipette with water and alcohol immediately after use and let it dry.
- Take care the suspension is not disturbed. Obtain the depth of sampling for the cutoff point of 0.002 mm at the suspension temperature (Table 2.1.1).
- Draw an aliquot from this depth from the surface of the suspension (which will be at a slightly lower level now) after 5 hours 30 minutes exactly in the same manner as described earlier.
- Transfer the aliquot to a previously weighed aluminium dish, evaporate to dryness and keep overnight in an oven at 105 °C.
- Cool in a desiccator and weigh quickly to the nearest mg. This weight (Z) is the weight of the aliquot of clay.
- Both the aliquots will contain a certain amount of dispersing agent which adds to the weight. To obtain a correction factor ($c$) for this weight dilute 10 mL of the dispersing agent to 1000 mL, transfer a similar aliquot, using the same pipette to a previously weighed aluminium dish, dry and weigh to the nearest mg.

- Transfer the dried sand to a nest of this first five sieves (see list of apparatus) placed in a flat porcelain basin. Wash the materials through the sieves using jet of water. Thus separation of various sand fractions is achieved.
- Wash each fraction into weighed dish, decant off the water, dry and weigh.
- Check the sum of the weights of the individual fractions against total weight of sand recorded earlier.

**Calculations**

$$\text{Clay + International (Fine) Silt (\%)} = \frac{(Y-c)}{X} \times \frac{1000}{v} \times 100$$

$$\text{Clay (\%)} = \left(\frac{(Z-c)}{X} \times \frac{1000}{v} \times 100\right)$$

where X = weight of treated oven dry soil (base weight), in gram

Y = weight of international silt (fine silt) plus clay fraction aliquot, in gram

Z = weight of clay fraction aliquot, in gram

$c$ = weight of dispersing agent in aliquot, in gram

$v$ = volume of aliquot, mL

% USDA Silt (0.05–0.002 mm fraction) = {100 – (%sand + %clay)}

Report the following fraction covering both the systems

| | | |
|---|---|---|
| USDA | Total sand | 2.0–0.05 mm |
| | Very coarse sand | 2.0–1.0 mm |
| | Coarse sand | 1.0–0.5 mm |
| | Medium sand | 0.5–0.25 mm |
| | Fine sand | 0.25–0.10 mm |
| | Very fine sand | 0.10–0.05 mm |
| | Silt | 0.05–0.002 mm |
| | Clay | < 0.002 mm |
| International | I Coarse sand | 2.0–0.20 mm |
| | II Fine sand | 0.20–0.02 mm |
| | III Silt | 0.02–0.002 mm |
| | IV Clay | < 0.002 mm |

Definition of clay is same in both the system. The 0.2 mm sieve is required to determine International I and II (coarse sand and fine sand).

Finally determine the textural class of the soil from the triangular chart of Soil Survey Staff (1951) of USDA (Fig. 2.1) (see Article 2.1.2 on determination of soil textural class).

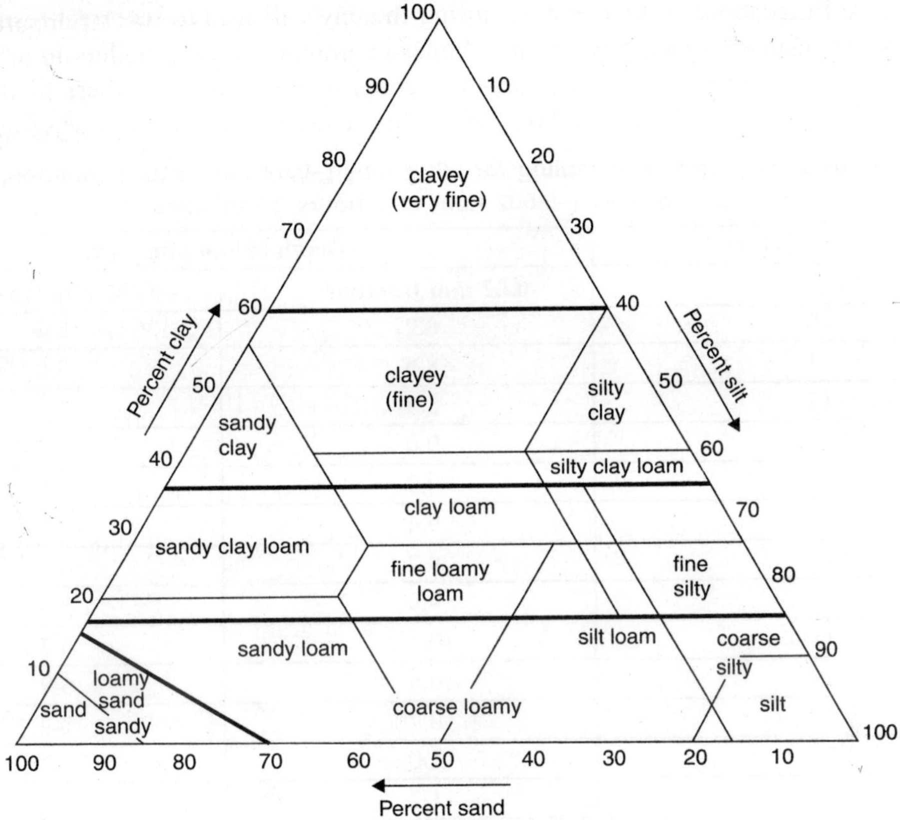


**Fig. 2.1** Triangular textural diagram U.S.D.A.

***Notes***

- Soils which are non calcareous and contains less than 0.5% organic carbon usually need no pretreatment prior to dispersion. Pretreatments for removal of calcium carbonate in case of highly calcareous soil and removal of free iron oxides in case of ferruginous soils, are optional and are thus not recommended in general practice. Data on particle size analysis in ferruginous soils which are available in literature have been generally obtained without pretreatment for removing free iron oxides. Saline soils may need special treatment according to the kind and amount of salts present. When acid is used, the subsequent filtration and washing with water removes soluble salts or reduce them to smaller amounts unlikely to affect the analysis otherwise the quantities of soluble salts will affect the weight of oven dry soil. On the other hand it must be kept in mind that washing the soils entirely free of salts may lead to deflocculation of clay and passage of the clay particles through the filter. Hence, washing must not be too prolonged except to reduce the amount of gypsum to amounts which will not interfere with proper dispersion of silt and clay particles.
- It is often claimed that, sodium hexameta phosphate is not effective in dispersing lateritic soils and soils containing much colloidal iron or aluminium oxides. Better dispersion may be obtained with sodium hydroxide, with ammonium carbonate and sodium hydroxide.

- Large fluctuations in temperature during the day will lead to incorrect results for clay. Use of insulating jacket for each cylinder or immersion of cylinders in a thermostatic trough will largely eliminate errors on account of temperature. If these facilities are not available a room of only slightly fluctuating temperature should be chosen.

**Table 2.1.1. Depth of sampling for silt + clay (–0.02 mm) after 5 minutes, and for clay (–0.002 mm) at 5 hours 30 minutes**

| Temperature | Depth of Sampling (cm) | |
|---|---|---|
| (°C) | –0.02 mm fraction | –0.002 mm fraction |
| 10 | 8.2 | 5.4 |
| 11 | 8.5 | 5.6 |
| 12 | 8.7 | 5.7 |
| 13 | 9.0 | 5.9 |
| 14 | 9.2 | 6.1 |
| 15 | 9.4 | 6.2 |
| 16 | 9.7 | 6.4 |
| 17 | 9.9 | 6.6 |
| 18 | 10.1 | 6.7 |
| 19 | 10.4 | 6.9 |
| 20 | 10.7 | 7.1 |
| 21 | 11.0 | 7.2 |
| 22 | 11.2 | 7.4 |
| 23 | 11.5 | 7.6 |
| 24 | 11.8 | 7.8 |
| 25 | 12.1 | 8.0 |
| 26 | 12.3 | 8.1 |
| 27 | 12.6 | 8.3 |
| 28 | 12.9 | 8.5 |
| 29 | 13.2 | 8.7 |
| 30 | 13.5 | 8.9 |
| 31 | 13.8 | 9.1 |
| 32 | 14.1 | 9.3 |
| 33 | 14.3 | 9.5 |
| 34 | 14.7 | 9.7 |
| 35 | 14.9 | 9.9 |
| 36 | 15.2 | 10.1 |
| 37 | 15.5 | 10.3 |
| 38 | 15.9 | 10.5 |
| 39 | 16.2 | 10.7 |
| 40 | 16.5 | 10.9 |

## 2.1.2 Hydrometer Method

### Principle

Hydrometer method has widespread use in routine work of particle size analysis where quick determinations are necessary and extreme accuracy is not required. Like International Pipette Method this method is also based on the principle of dispersion and sedimentation techniques employed to a given weight of soil sample. Sedimentation refers to the settling rates of the dispersed particles in water, which is function of particle size and is governed by Stoke's law (eq. 2.1.7). Theoretically the hydrometer measures the density of soil suspension. In practice, an average density to the depth of the inserted hydrometer is taken. The hydrometer is based on the fact that the suspension at a given depth decreases as an initially homogeneous dispersed suspension settles. The rate of decrease in density, at any given depth, is directly related to the settling velocities of the particles, which in turn are related to their size. The hydrometer reading indicates that 4 minutes after sedimentation particles greater than 0.02 mm settle, while after 2 hours, particles of size less than 0.002 mm are left in the soil suspension.

### Equipment and apparatus

- Standard Hydrometer with Bouyoucos scale in $gl^{-1}$
- Electrically given mixer
- Dispersing/Stirring cup
- Graduated cylinder (1000 mL) with rubber stopper
- Thermometer
- Stopwatch
- Hot plate
- Beaker and watch glass

### Reagents

- 5% sodium hexametaphosphate, 50 g calgon/litre.
- 6% $H_2O_2$

### Procedure

- Weigh 50 g fine textured soil or 100 g coarse textured soil (>75–80% sand) which have been passed through a 2 mm sieve based on oven dry condition into a beaker.
- Add 50 mL of 6% $H_2O_2$ and cover the beaker with a watch glass and place it on a water bath until oxidation of organic matter is complete (indicated by the presence of effervescence). Remove the beaker and cool.
- After cessation of frothing transfer the contents into a dispersing cup with about 400 mL of distilled water.
- Add to it 100 mL of calgon solution.
- Stir the suspension for 10 minutes by an electric stirrer.
- Transfer the suspension into a litre graduated cylinder and make up the suspension up to 1 litre mark with distilled water.

- Stopper the mouth of the cylinder and shake vigorously upside down and back several times for about 1 minute.
- Place the cylinder on a table and note the time immediately.
- Dip the hydrometer into the suspension and take the first reading after 4 minutes when particles > 0.02 mm have settled (Start inserting the hydrometer 10 seconds in advance of the reading time).
- Carefully remove the hydrometer and wash with distilled water and note down the temperature of the suspension.

***Note:*** The hydrometer is calibrated at 67 °F (19.4 °C). If the suspension temperature is above 67 °F, the correction is added, and if below, the correction is subtracted. The correction is equal to the difference between the experimental temperature and 67 °F, multiplied by 0.2.

For conversion of °F to °C the following equation is used $\frac{C}{5} = \frac{F-32}{9}$

- Keep the suspension undisturbed and dip the hydrometer again at the end of 2 hours after initial shaking was stopped. Now, the particles greater than 0.002 mm (sand plus silt) have settled. Record the hydrometer reading.

Calculate the percentage of sand, silt and clay and determine the textural class using ISSS textural triangle.

## Calculations

Let, 4 minute hydrometer reading be $x$ at 77 °F, when 50 g oven dry sample was used.

Corrected hydrometer reading = $[x + (77 - 67) \times 0.2]$ = Y say

% (silt + clay) in the suspension = $\left(\frac{Y}{50} \times 100\right)$

Now let wt. of soil = $a$ gram

Hydrometer reading at 4 min = $b$,

Working temperature for 4 minute observation = $c$

Corrected hydrometer reading at 4 minutes = $d$

Hydrometer reading at 2 hours = $e$

Working temperature for 2 hours observation = $f$

Corrected hydrometer reading at 2 hours = $g$

Amount of silt plus clay = $d$ gram

Amount of clay = $g$ gram

Amount of silt = $(d-g)$ gram

$$\% \text{ clay} = \frac{g}{a} \times 100$$

$$\% \text{ silt} = \frac{d-g}{a} \times 100$$

$$\% \text{ sand} = [100 - (\% \text{ silt} + \% \text{ clay})]$$

It is to be kept in mind that the shortcoming of this method is the lack of any account of the content of organic matter of soil.

### Determination of soil textural class

The texture of soil is determined from the relative proportion of sand, silt and clay. Two systems of soil texture classifications, as suggested by ISSS and USDA, are in common use. Both the system make use of an equilateral triangle whose area is divided into 12 compartments, each representing a textural class. The difference is primarily due to differences in size ranges of sand and silt fractions. The triangle based on ISSS size fraction is given in Fig. 2.2. For the determination of soil textural class, locate the clay and silt percentages on the respective sides of the triangle. Draw a line inward parallel to the sand axis in the former case and parallel to clay axis in the latter case. The compartment in which the two lines intersect in the texture of the soil.

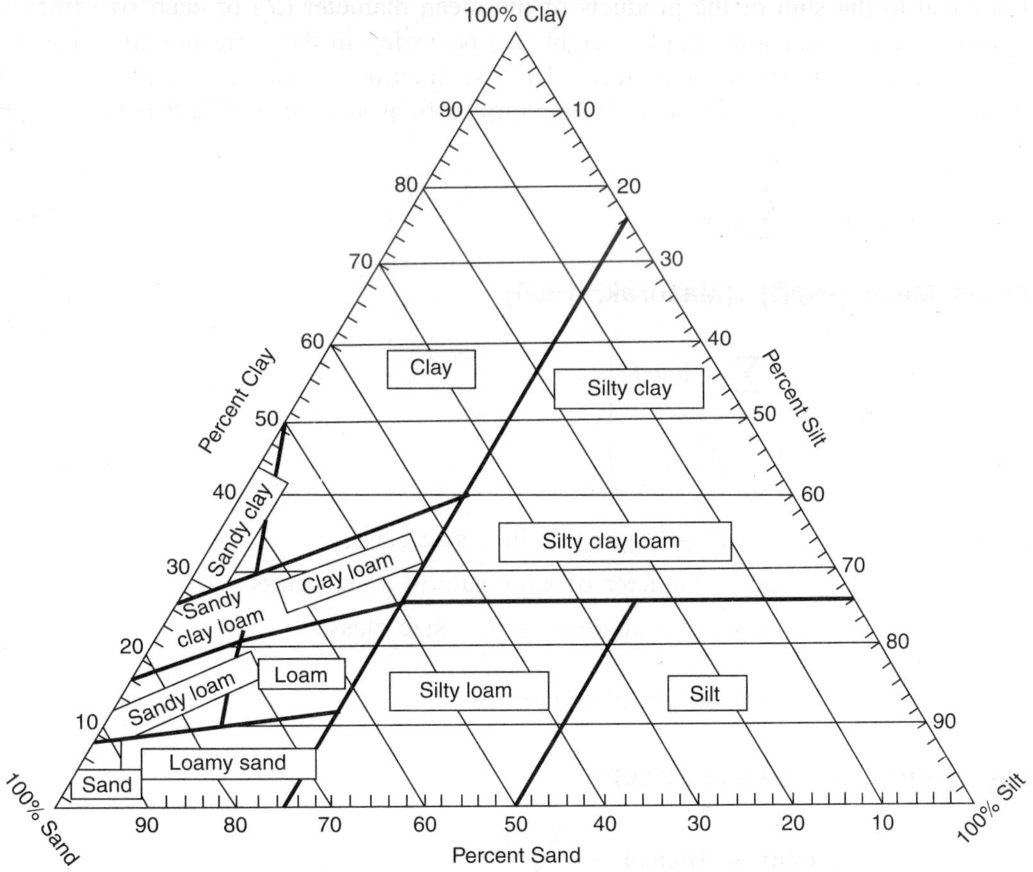


**Fig. 2.2** Triangular textural diagram based on ISSS classification

## 2.2 AGGREGATE SIZE ANALYSIS BY WET SIEVING METHOD

### Principle

A known amount of soil sample collected from the field is immersed in water for a short period of time for wetting. The wetted sample is sieved through a nest of sieves in Yoder's apparatus, which raises and lowers the nest of sieves. This is done to simulate the disruptive forces of water and facilitate the sieving of water stable aggregates through sieves. After about 30 minutes, sieves along with aggregates are removed and dried in an oven at 105°C. The dry aggregates are collected and weighed. The weight of these aggregates also includes weight of primary particles of respective sieve sizes. Hence, soil of these aggregates from sieves are dispersed and passed through the respective sieves. The mass retained on respective sieves represents primary particles or sand fraction and therefore is subtracted from the mass of aggregates to get correct estimate of the aggregates. To represent the aggregation status of soil by a single value, following indices are evaluated:

**Mean Weight Diameter (MWD): (van Bavel, 1949)**

MWD is equal to the sum of the products of the mean diameter ($d_i$) of each size fraction and the proportion of the total soil sample weight ($w_i$) occurring in the corresponding size fraction, where the summation is carried out over all $n$ size fractions including the one which passes through the finest sieve. [The entire soil sample must be passed through a 8 mm sieve prior to analysis].

$$\text{MWD} = \sum_{i=1}^{n} d_i w_i$$

**Geometric Mean (GMD): (Mazurak, 1950)**

$$\text{GMD} = \left( \frac{\sum_{i=1}^{n} w_i \log d_i}{\sum_{i=1}^{n} w_i} \right)$$

where $w_i$ = weight of aggregate in a size class,

$d_i$ = mean diameter of aggregate in a size class,

$n$ = number of aggregates in a size class,

$\sum_{i=1}^{n} w_i$ = total weight of sample.

**Stability coefficient: Russel (1938)**

$$\text{Stability coefficient} = \frac{X - Y}{Y}$$

where, X = percent of primary particles < 0.25 mm in diameter obtained from particle size distribution analysis.

Y = percent of soil particles < 0.25 mm in diameter obtained by wet sieve aggregate analysis method.

## Stability index: Alderfer and Morkle (1941)

Stability index (S.I.) is usually evaluated by measuring the area between the curves of aggregate distribution and primary particles distribution on the coarser side of intersection (Fig. 2.3a).

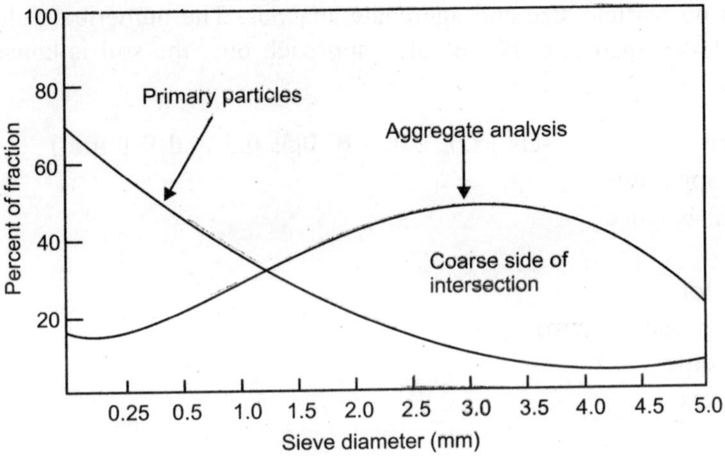


**Fig. 2.3a** Curve of aggregate and primary particles distributions

$$\text{S.I.} = \frac{\text{Percent of fraction}}{\text{Area of coarse side of intersection (measured by planimeter)}}$$

The value is always lesser than unity.

## Aggregate index: Van Bavel (1953)

Aggregate Index (A.I.) is obtained by measuring the area between the two curves obtained by plotting the percentage of soil particles below a size range against that size class (upper limit) and other by plotting the percentages of primary particles below the size range against the respective size classes on the same graph (Fig. 2.3b).

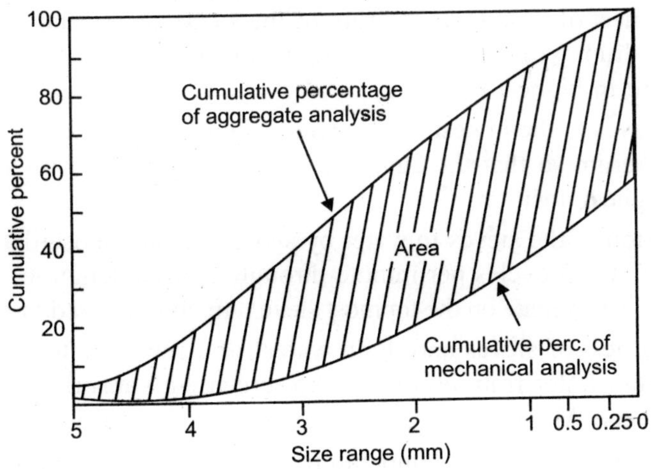


**Fig. 2.3b** Curves of cumulative percentage of aggregate and partical distribution analyses

$$\text{A.I.} = \frac{\text{Cumulative percentage} = 100}{\text{Area between the two curves}}$$

***Note:*** All the three indices viz. the stability coefficient, the stability index and the aggregate index can be evaluated by particle size and aggregate analysis. The numerical values of all the indices must always be lesser than one. If the value approach one, the soil is considered to be good.

## Equipments

- Standard sieves – 2 sets (5.0, 2.0, 1.0, 0.5, 0.2 and 0.1 mm)
- Yoder apparatus
- Physical balance
- Oven
- Desiccator
- Watch glasses (8 cm)
- Wash bottle
- Can boxes

## Procedure

- Take 250 g of air dry solid clods (approx) and remove large gravels or roots. Break them into smaller aggregates with hand in such a way that they pass through 8 mm screen and are retained on 5 mm screen.
- Weigh 50 g aggregates (5–8 mm) in three watch glasses. Keep one such sample in an oven at 105 °C for moisture determination and use the remaining two for analysis in duplicate.
- Arrange two sets of six sieves viz. 5, 2, 1, 0.5, 0.2 and 0.1 mm in such a way that the uppermost sieve has the largest mesh size and the sieve at the bottom should have smallest mesh size.

## Aggregate sample

Spread the aggregate sample uniformly on the top sieve and add 10 mL of salt free water. After 5 minutes, spray another 5–100 mL of water and wait for 3–5 minutes.

- Transfer the nest of sieves to the drum of the sieve. Shake and clamp them in position. Then fill the drum with salt free water up to a level, slightly below the top screen, when the sieves are in highest position.
- Lower the sieves to the lowest position and wet the aggregates for 10 minutes. Fill more water in the drum so that the aggregates are just covered with water when the sieves are in the highest position.
- Start oscillation of the sieves in water by switching on the oscillator for 30 minutes at a frequency of 30–35 cycles per minute through a stroke length of about 3.5 cm. Ensure that the sample aggregate on the topmost sieve remain immersed throughout the full stroke.
- Take out the nest of sieves and drain water for a few minutes in an inclined position. Remove excess water from bottom of the screen with absorbent tissue paper and place on paper sheets. Allow the aggregates on each sieve to dry and harden in air.
- Dry the soil in an oven at a temperature not exceeding 75 °C, since high temperature results in adherence of some soils to the sieves. When dry (usually after 30–40 minutes)

transfer the soil from each sieve separately into can boxes, dry overnight in an oven at 105° and weigh.

### Dispersed sample

For the purpose of estimation how much of the soil, retained on the sieves, represents aggregates and how much is gravel or sand, transfer the aggregates of each to 250 mL beakers separately and disperse them with $H_2O_2$ and HCl treatments. Pass the dispersed aggregates again through the same sieves in which they were previously retained. Collect the unaggregated primary particles from each sieve, in can boxes according to procedure already described and record their oven dry weight. Now calculate the percentage of aggregated soil particles on different sieves. Plot a graph between the accumulated percentage of soil remaining on each sieve as ordinate and the upper limit of each size fraction as abcissa. From the graph evaluate the mean weight diameter (MWD) of aggregates by measuring the area under the curve. Also find out the MWD in mm by computation.

## 2.3 PARTICLE DENSITY ($D_p$)

Particle density ($D_p$) is usually defined as mass (weight) per unit volume of soil solids. It is generally expressed in g/cm. For mineral soils the value usually varies between 2.60 and 2.75 g/cm.

### Principle

A given amount of dry soil when immersed in a definite volume of water, expels air and results in displacement of an equal volume of water. The volume of soil particles is determined by measuring the volume of water displaced in the pycnometer bottle.

### Equipment

- A Pycnometer
- Pipette (20 mL capacity)
- An analytical balance
- Hot plate or water bath

### Procedure

- Fill up a dry and clean pycnometer with deaerated water. Note its temperature.
- Wipe out the surface of the pycnometer and replace the stopper and weight.
- Empty the pycnometer and fill into it 10 g oven-dried soil.
- Fill the pycnometer to about half with water using the pipette and wash with a jet of water any soil particles sticking to the inner side of the neck.
- Expel the entrapped air by gently boiling the contents.
- Allow the contents to cool to room temperature and fill the pycnometer to the brim with boiled and cooled water.
- Fix the stopper and clean the outerside of pycnometer with water and weight it.

### Calculations

$$\text{Wt. of water filled pycnometer} = W_{pw} \text{ gram}$$
$$\text{Wt. of dry soil} = 10 \text{ gram}$$
$$\text{Wt. of pycnometer + water + soil} = W_{psw} \text{ gram}$$
$$\text{Volume of water displaced (volume of soil solids)} = (W_{pw} + 10 - W_{psw}) \text{ cm}^3$$
$$\text{Particle density of soil} = (10/W_{pw} + 10 - W_{psw}) \text{ g/cm}^3$$

## 2.4 BULK DENSITY ($D_b$)

Bulk density ($D_b$) is defined as the mass (weight) of a unit volume of dry soil. This volume includes both solid and pores. The values for clay, clay loam and silt loam surface soils varies from 1.00 to 1.60 g/cm$^3$, for sand and sandy loams from 1.20 to 1,80 g/cm$^3$. Fine-texture soils tend to have lower bulk densities and therefore higher porosities in comparison to the coarse textured soils due to loose packing of the clay particles. Bulk density measurement for soils is important since it determines the degree of compactness as a measure for soil structure and is used for calculating pore space of soils.

### 2.4.1 Core Sampler Method (Bulk Density of Undisturbed Soil)

#### Principle

The method advocates sampling a soil core *in situ* using core sampler (cylindrical metal sampler) from a desired depth and determining the mass of solids together with the water content of the core. **For** this purpose first the wet core is weighed and then dried to a constant weight in an oven at 105°C (24 hours may be required for drying) and then re-weighing after cooling. Subsequently bulk density is calculated from the measurement of bulk volume, using the core length and the diameter of the cutting edge of the sampler.

#### Apparatus

Core sampler; vernier slide caliper; can boxes; oven balance; desiccator; knife spatula.

#### Procedure

- Push the core sampler vertically into level surface deep enough to fill the sampler can in the sampler.
- With the help of spade, dig out the sampler and remove the sample can without disturbing the soil core contained therein.
- Remove off the extra soil from both ends of the sample can by levelling with a sharp knife.
- Weigh the sample can along with the soil
- Take out the sample and weigh the can again
- Place a part of the moist soil in a can box, and find out water content by drying it in an oven at 105°C.

- Determine the length and inner diameter of the sample can.
  Volume of the soil (whole core) = inside volume of the can = $\pi r^2 l$
  where $r$ = radius and $l$ = length in cm.

### Observations and Calculations

$$\text{Wt. of can + soil} = W_1 \text{ gram}$$

$$\text{Wt. of can} = W_2 \text{ gram}$$

$$\text{Wt. of can box} = W_3 \text{ gram}$$

$$\text{Wt. of can box + moist soil} = W_4 \text{ gram}$$

$$\text{Wt. of can box + dry soil} = W_5 \text{ gram}$$

$$\therefore \quad \text{Wt. of oven dry soil} = (W_5 - W_3) \text{ gram} = x \text{ gram (say)}$$

$$\text{Wt. of water in soil} = (W_4 - W_3) \text{ gram} = y \text{ gram (say)}$$

$$\text{Wt. of the moist soil (whole core)} = (W_1 - W_2) \text{ gram} = z \text{ gram (say)}$$

$$\text{Water content in the soil} = \left(\frac{y}{x} \times 100\right) = p\% \text{ (say)}$$

$$\text{Wt. of the dry soil (whole core)} = \left(\frac{z \times 100}{p + 100}\right) = q \text{ gram (say)}$$

$$\text{Bulk density of the soil} = \frac{q}{\pi r^2 l} \text{ g cm}^{-3}.$$

### 2.4.2 Clod Saturation Method

#### Principle

The volume of water absorbed on saturating a dry clod is equal to the volume of pores. The volume of solid phase of soil is determined by the ratio of mass of solids and particle density. The bulk density is determined by dividing the mass of clod by its total volume.

The bulk volume ($V_b$) of a dry clod weighing ($M_s$) may be expressed by the equation

$$V_b = V_s + V_p \qquad \text{...(2.4.2.1)}$$

where $V_s$ is the volume of soil solids and $V_p$ is the volume of pores.

$V_s$ is calculated as:

$$V_s = \frac{M_s}{D_p} \qquad \text{...(2.4.2.2)}$$

where $M_s$ is the mass of dry soil and $D_p$ is the particle density.

$D_p$ may either be determined or an average value of 2.65 g/cm$^3$ can be used when determination is not possible.

$$\text{Now, } V_b = \frac{M_s}{D_p} + V_p \qquad \text{...(2.4.2.3)}$$

The value of $V_p$ is determined by difference in mass of saturated and the dry clod which equals to the mass (or volume) of water adsorbed, $V_{wa}$.

$$\text{Hence, } V_b = \frac{M_s}{D_p} + V_{wa} \qquad \text{...(2.4.2.4)}$$

Since bulk density is mass per unit bulk volume of the soil,

$$\text{Therefore, } D_b = \frac{M_s}{V_b} \qquad \text{...(2.4.2.5)}$$

## Equipment

- A balance
- Sand column
- Watch glass
- Filter paper

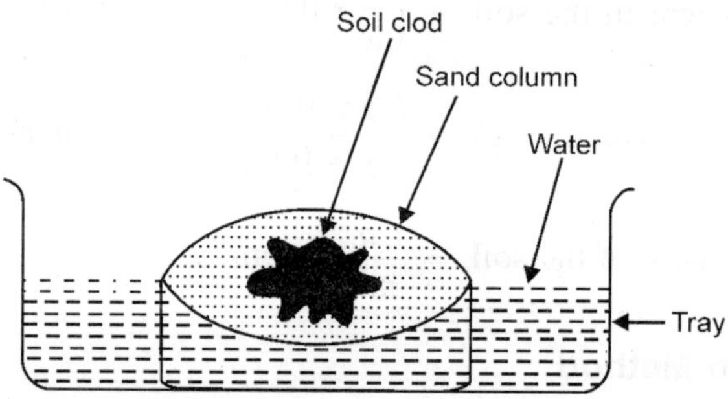


**Fig. 2.4** Sand column with a soil clod on it for saturation

## Procedure

- Take an oven dry clod and weigh it.
- Saturate the clod by capillarity, by placing it on a filter paper disc placed on the sand column. The clod glistens upon saturation.
- After saturation of the clod, transfer the clod along with filter paper disc to a watch glass using a spatula and weigh immediately.
- Determine the mass of saturated clod by subtracting the mass of watch glass and a wetted filter paper disc of the size placed below the clod.
- Determine the mass of water absorbed by the clod.

## Calculations

Mass of dry clod = $M_s$ gram

Particle density of soil = $D_p$ g/cm$^3$

Volume of soil solids = $V_s = \dfrac{M_s}{D_p}$ cm$^3$

Mass of saturated clod = $M_{sm}$ gram

Mass of water absorbed during saturation = $(M_{sm} - M_s)$ gram

$$\text{Density of water } (D_w) = 1 \text{ g/cm}^3$$

$$\text{Volume of water absorbed } (V_{wa}) = [(M_{sm} - M_s)/D_w] \text{ cm}^3$$

$$\text{Volume of clod } (V_b) = (V_s + V_{wa}) \text{ cm}^3$$

$$\text{Bulk density} = \frac{M_s}{V_b} \text{ g/cm}^3$$

## 2.5 TOTAL POROSITY

Porosity of a soil sample is the volume which is occupied by air and water. Mathematically, it is the ratio of volume of pore space to total volume of soil. Porosity is governed by the arrangement or orientation of the solid particles. Total porosity gives an idea only about the total storage capacity of soil for fluids or gases.

The volume percentage of the total soil bulk not occupied by the solid particles usually it is expressed as

$$\% \text{ pore space} = \left[100 - \frac{(D_b \times 100)}{D_p}\right]$$

This is obtained as follows:

By definition

$$D_p = \frac{W_s}{V_s} \quad \text{...(2.5.1)}$$

$$D_b = \frac{W_s}{V_s + V_p} \quad \text{...(2.5.2)}$$

where
$W_s$ = weight of soil solids,
$V_s$ = volume of solid,
$V_p$ = volume of pores,
$V_s + V_p$ = total soil volume,
$D_b$ = bulk density,
$D_p$ = particle density,

$$\therefore \quad W_s = D_p V_s \quad \text{...(2.5.3)}$$

$$\text{and} \quad W_s = D_b (V_s + V_p) \quad \text{...(2.5.4)}$$

$$\text{Hence} \quad D_p \cdot V_s = D_b (V_s + V_p) \quad \text{...(2.5.5)}$$

$$\text{or} \quad \frac{D_b}{D_p} = \frac{V_s}{V_s + V_p} \quad \text{...(2.5.6)}$$

It can be written that $\% \text{ solid space} = \frac{D_b}{D_p} \times 100$ ...(2.5.7)

Also, since (% pore space + % solid space) = 100 ...(2.5.8)

% pore space =100 – % solid space ...(2.5.9)

$$= 100 - \frac{D_b}{D_p} \times 100 \quad ...(2.5.10)$$

**Procedure**

- Determine the soil bulk density ($D_b$) and particle density ($D_p$)
- Calculate the total porosity by the equation,

$$\text{Total porosity} \quad (f) = 1 - \frac{D_b}{D_p} \quad ...(2.5.11)$$

## 2.6 AIR-FILLED POROSITY

### 2.6.1 Difference Method

**Principle**

Air-filled porosity of the soil at a given soil water content is obtained by subtracting the water content value from its total pore space.

**Procedure**

- Determine water content of a soil sample gravimetrically
- Convert the soil water content on mass basis to volume basis by multiplying the former with bulk density
- Calculate total porosity from the values of soil bulk density and particle density
- Subtract the value of volumetric water content from the value of total porosity to get the value of air-filled porosity

**Calculations**

$$\text{Bulk density of soil} = D_b \text{ g/cm}^3$$

$$\text{Particle density of soil} = D_p \text{ g/cm}^3$$

$$\text{Total porosity } (f) = 1 - \frac{D_b}{D_p}$$

$$\text{Volumetric water content} = \theta$$

$$\text{Air porosity} = (f - \theta)$$

### 2.6.2 Air Pycnometer Method

**Principle**

The mean advantage of air-pycnometer over the weight difference method is the speed with which the measurements can be made. A single determination of the volume of the soil and water in a given sample at any moisture content may be made in less than two minutes. The volume occupied by air in the sample may then be determined by simply subtracting this volume from the total volume of the soil samples cylinder.

The procedure of air-pycnometer is based on Boyle's law: $P_1V_1 = P_2V_2$ in which P and V are gas pressure and volume respectively at a particular temperature. The volume of air space in a sample is measured by observing the resulting pressure when a known volume of gas at a known pressure expands into a larger volume that includes the air space in the sample.

The principle is very simple as demonstrated in (Fig. 2.5) where two vessels (containing air) A and B are connected through a valve E. Volume and pressure of vessels A and B are $V_1P_1$ and $V_2P_2$ respectively. If the valve E is opened the pressure in A and B will be equal. Let it be $P_3$.

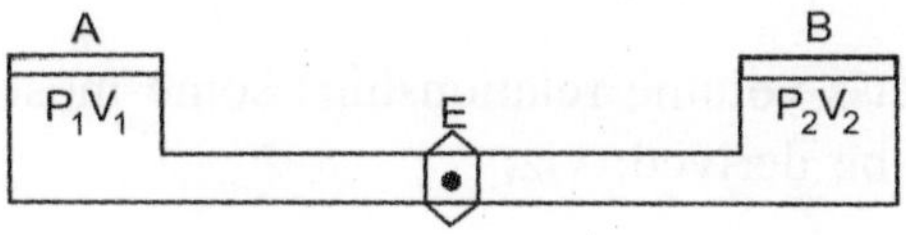


**Fig. 2.5** Air pycnometer

Therefore,

$$P_1V_1 + P_2V_2 = P_3V_1 + P_3V_2$$

$$P_3(V_1 + V_2) = P_1V_1 + P_2V_2$$

$$P_3 = \left(\frac{P_1V_1 + P_2V_2}{V_1 + V_2}\right) \quad ...(2.6.2.1)$$

In case,

$$V_1 = V_2$$

then,

$$P_3 = \frac{P_1V_1 + P_2V_1}{V_1 + V_1} = V_1\frac{V_1(P_1 + P_2)}{2V_1} = \frac{P_1 + P_2}{2} \quad ...(2.6.2.2)$$

## Apparatus

- Air pycnometer (Fig. 2.5)
- Metal plates of sample chamber size.

## Procedure

- Insert enough metal plates into the sample chamber to occupy 50% of its volume.
- Inflate the reservoir to 5 psi (pounds per square inch) on the gauge.
- Open the connecting valve.
- Record the pressure when flow has stopped *i.e.,* when equilibrium is attained.
- Repeat the above steps when various portions 60%, 70%, 80%, 90% and 100% (*i.e.,* 40%, 30%, 20%, 10% and 0% air porosity) of the sample chamber are occupied by metal plates.
- Construct a graph relating final pressure to volume of air space in the sample chamber.

## Measurements

- Insert the soil sample into the sample chamber.
- Inflate the reservoir to 5 psi.

- Open the connecting valve.
- Record the final pressure.
- Using the calibration curve already prepared, find air space against final pressure.

**Calculations**

Final air pressure in the reservoir with soil sample = $P_1$

Percent air-space corresponding to observed $P_1$ (from calibration curve) = $f_a$

### 2.6.3 Interrelations

From basic definitions of mass-volume relationships some most useful interrelations amongst various parameters can also be derived; viz.

- Relation between porosity ($f$) and void ratio ($e$)

$$e = \frac{f}{(1-f)} \quad \text{...(2.6.3.1)}$$

- Relation between volume wetness ($\theta$) and degree of wetness ($s$)

$$s = \frac{\theta}{f} \quad \text{...(2.6.3.2)}$$

- Relation between porosity ($f$) and bulk density ($D_b$)

$$f = \left(1 - \frac{D_b}{D_p}\right) \quad \text{...(2.6.3.3)}$$

- Relation between mass wetness (W) and volume wetness ($\theta$)

$$\theta = W \frac{D_b}{D_w} \text{ ($D_w$ is density of water = 1 g cm}^{-3}\text{)} \quad \text{...(2.6.3.4)}$$

- Relation between volume wetness ($\theta$), air-filled porosity ($f_a$) and degree of saturation ($s$)

$$f_a = f - \theta = f(1 - s) \quad \text{...(2.6.3.5)}$$

$$\theta = f - f_a \quad \text{...(2.6.3.6)}$$

- Relation between specific volume ($S_v$) and bulk density ($D_b$)

$$S_v = \frac{1}{D_b} \quad \text{...(2.6.3.7)}$$

## 2.7 TOTAL SURFACE AREA DETERMINATION OF SOIL BY ETHYLENE GLYCOL EQUILIBRATION METHOD

Total surface area of a soil is defined as area per unit weight of clay or soil and is expressed as ($m^2g^{-1}$). The fact that ethylene glycol forms a monomolecular layer on the clay surface, forms the very basis of specific area measurement. It is already established that to form a monolayer on each square meter of clay surface 0.00031g of ethylene glycol is required. (Dyal and Hendricks, 1950). A knowledge of specific surface area is essential for the determination of surface charge density of solids or clays for predicting the saturation percentage of the exchanger with monovalent cation.

## Principle

Solid materials adsorb a mono-molecular layer of adsorbate at a given temperature and pressure. A knowledge of the molecular size (diameter) and mass of the adsorbate adsorbed enables one to calculate the specific surface area.

## Equipment, Apparatus and Reagents

- Glass dishes with lid
- Vacuum desiccator
- Pipette 1 mL
- Electrical balance
- Petridishes
- Ethylene glycol
- $CaCl_2$ – glycol solvate buffer (Dry completely about 120 g of 40-mesh $CaCl_2$ in an oven at 210°C. Weigh 20 g of glycol in 400 mL of a pyrex beaker and add 100 g of dry $CaCl_2$ to it without cooling. Mix the contents thoroughly with spatula. Spread the solvate uniformly in culture chamber for cooling and store in a sealed desiccator).

## Procedure

- Keep the soil sample in a vacuum desiccator with $P_2O_5$ or $CaCl_2$ and evacuate for 5 to 6 hours to dry completely.
- Weigh accurately 1.1g soil or 0.3g clay in two dishes and spread evenly.
- Wet one sample completely with minimum (1 mL or less) ethylene glycol by adding dropwise from 1 mL pipette.
- Enclose the wetted and non wetted samples and 120 g $CaCl_2$ – glycol solvate in a chamber to minimize the diffusion path of glycol vapour and place them in a vacuum desiccator.
- Subject to vacuum for 48 hours with 2–3 evacuations for about 30 to 60 minutes after 16 to 24 hours.
- Release the vacuum, cover the samples and weigh accurately.
- Average the weight of the ethylene glycol, retained by 1g of the sample and obtain specific surface area per gram of the sample in $m^2/g$ upon division by 0.00031.

## Calculations

(Perform both for glycol wetted and non-wetted samples)

$$\text{Wt. of dish} = a \text{ gram}$$

$$\text{Wt. of dish + soil} = b \text{ gram}$$

Wt. of soil $= (b - a)$ gram $= c$ gram

Wt. of dish + soil + ethylene glycol $= d$ gram

Wt. of glycol retained $= (d - c)$ gram $= e$ gram

$$\text{Specific surface of soil} = \frac{e}{c \times 0.00031} = f\ \text{m}^2\text{g}^{-1}$$

$$\text{Corrected specific surface of soil} = [f(\text{I}) - f(\text{II})]\ \text{m}^2\text{g}^{-1}$$

## 2.8 DETERMINATION OF HEIGHT OF CAPILLARY RISE OF WATER IN SOIL

Capillary rise of water in soils is of great agricultural significance when movement of salts is considered in soils; where the dissolved, salts present in soil solution move upward to the surface by capillarity. As a matter of fact, capillary rise of salts from shallow ground water table is the cause or surface salinization. When groundwater does not contain appreciable amounts of salts capillary rise proves to be beneficial to crop growth in a way to meet a part of water requirement of crops as for example, in shallow water table areas of Tarai soils in Uttar Pradesh. Water pressure in the capillary is negative and water is thus said to be held under tension or suction. Water movement in capillaries results due to pressure difference between adjacent capillaries. Water in fact, moves from high pressure zone to low pressure zone *i.e.,* from large capillaries to small capillaries. In soil system the connected pore space act as capillary tubes. A knowledge of capillary rise phenomenon in soils is helpful in a way to decide the depth to which water table should be lowered during soil reclamation and in maintaining favourable air-water regime for crop growth.

### Principle

Water forms a concave meniscus around the perimeter of a capillary tube due to the adsorption forces between the tube surface and the liquid as well as the cohesive forces from the liquid surface, known as 'surface tension'. The adhesion between the water molecules and glass surface tends to make the water move upward along the capillary sides and surface of water in the capillary concaves. Surface tension, on the contrary tends to level the surface in order to decrease the surface area. Simultaneous and repeated operation of these two phenomena results in capillary rise of water to a maximum height at which weight of water column equals upward component (cosine component) of surface tension force. The surface tension force acts all along the circumference ($2\pi r$) and tangential to the surface and is inclined at an angle $\theta$ (Fig. 2.6).

The upper meniscus is concave and if $\theta$ be the angle of contact then the vertical component of surface tension ($\gamma$) will be $\gamma \cos \theta$. The contact line of the meniscus with the wall of the tube is $2\pi r$. Hence net upward pull is ($2\pi r\, \gamma \cos \theta$). This is balanced by the weight of the liquid which has been drawn up. The weight of the liquid column is $(\pi r^2 h + v)\, \rho g$ where $v$ is the volume of the liquid in the curved meniscus itself, and $\rho$ is the density of the liquid.

Then,
$$2\pi r\, \gamma \cos \theta = (\pi r^2 h + v)\, \rho g \qquad ...(2.8.1)$$

The radius of curvature of the concave meniscus may be taken to be the same as the radius of capillary tube.

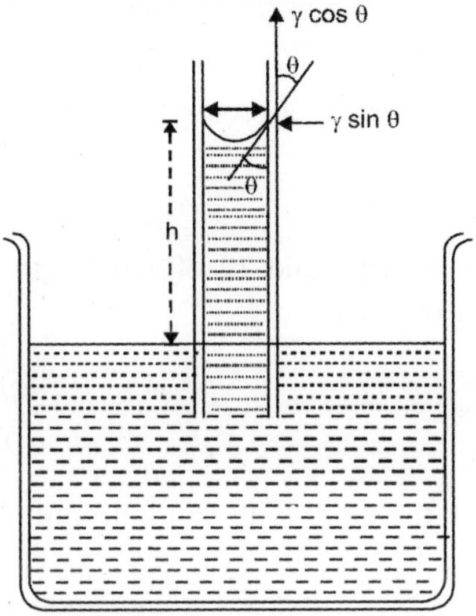


**Fig. 2.6** Capillary rise of water

Also, $$v = \pi r^3 - \frac{2}{3}\pi r^3 = \frac{1}{3}\pi r^3 \quad ...(2.8.2)$$

Hence $$2\pi r\ \gamma\ \cos\theta = (\pi r^2 h + \frac{1}{3}\pi r^3)\ \rho g \quad ...(2.8.3)$$

If the term $v$ in equation (2.8.1) is negligibly small, then

$$2\pi r\ \gamma\ \cos\theta = \pi r^2 h\ \rho g \quad ...(2.8.4)$$

$$h = \frac{2\gamma r\ g\cos\theta}{\pi r^2\ \rho g} \quad ...(2.8.5)$$

where $h$ = height of water (cm),
$r$ = radius of the tube (cm),
$\rho$ = density of water (g $cm^{-3}$),
$\gamma$ = surface tension (dynes $cm^{-1}$),
$\theta$ = contact angle,
$g$ = acceleration due to gravity (cm $sec^{-2}$).

For water which wets glass and also soil θ is acute or may be zero, so that cos θ is positive or 1. This implies that $h$ is also positive *i.e.,* the water in the capillary will rise above the water table outside the tube.

Capillary rise of water in soil simulates the condition of a capillary tube. Substituting,

$\gamma$ = 72 dyne $cm^{-1}$ at 25°C
$\rho$ = 1 g $cm^{-3}$
$g$ = 980 cm $sec^{-2}$
$\theta$ = 0

We get $$h = \frac{0.15}{r} \quad ...(2.8.6)$$

This relation shows inverse variation between $r$ and $h$ and that $rh$ is constant at a particular temperature.

## Apparatus

- Glass tubes of 2–3 cm diameter and 70–80 cm length with a scale in cm made from paper strips and pasted on them.
- A water trough
- Stand for holding tubes
- Spoon
- Rubber hammer to pack the tubes with soil uniformly
- Cheese cloth and string.

## Procedure

- Tie with string firmly the cheese cloth over the bottom of glass tubes.
- Pack the soil of different texture in some tubes (except two tubes) with the spoon, gently tapping the sides by the rubber hammer and ensure compact filling such that homogeneous soil profile is simulated.
- Of the two tubes, fill the lower half of one tube with one soil and top with other, while reversing the order in other tube. Take proper care that there is an abrupt boundary between the two textural distribution in one tube so as to simulate the field condition.
- Place a piece of filter paper at the top of the tubes and dip their lower ends in water and support the tubes.
- Record the height of water rise in the tubes at suitable time intervals (varying from 10 minutes to several hours and to days). Note down the time and date with each reading.

## Observations and calculations

- Record the results in the following tabular form.

| Sl. No. | Date/time of observation | Cumulative time from zero hour | Cumulative height of water rise (cm) | | | |
|---|---|---|---|---|---|---|
| | | | I | II | III | IV |
| 1 | | | | | | |
| 2 | | | | | | |

- Plot height of water rise versus time.
- Calculate the average pore size from the height of water rise in columns of homogeneous and layered soils.

## 2.9 DETERMINATION OF 'SINGLE VALUE PHYSICAL CONSTANTS' OF SOIL BY KEEN RACZ KOWSKI BOX MEASUREMENT

### Theory

The following soil physical characteristics are observed in this experiment:

- Apparent density
- Absolute specific gravity
- Maximum water holding capacity
- Percentage of pore space
- Volume expansion of 100 mL soil

### Equipment

KEEN BOX, brass box (5 cm diameter and height 1.6 cm approx.) with a perforated base just large enough in diameter to allow the cylinder to fit as tightly as possible.

### Procedure

- Weigh the keen box fitted up with a filter paper on a physical balance.
- Pack the box with air dry soil sample passed through 0.5 mm sieve by adding small quantities at a time and tapping the box after each addition to ensure even packing.
- Continue adding and tapping until the box is nearly full.
- Then add enough sample to fill the box.
- Strike off the surplus soil with a sharp blade or spatula to bring the soil in the box in level with the top of the box.
- Tap again and if necessary add soil, and level off once again.
- Weigh the box with air dry soil.
- Place the box with the soil in a tray or petridish containing distilled water to a depth of ¼ inch and leave overnight.
- After equilibrium is reached, remove the box from the petridish.
- Drain excess water by placing the box on table.
- Wipe the outside of the box with a dry towel and weigh immediately. Record the weight of keen box plus filter paper and saturated soil.
- Cut the expanded portion of the soil above the top of the box with a sharp knife into a previously weighed watch glass.
- Weigh the keen box with saturated residual soil.
- Weigh also the watch glass containing the surplus saturated soil.
- Dry both the box and the watch glass with saturated soil, in an oven at 105°C to a constant weight.

**From the set of readings calculate the**

- Apparent density
- Absolute specific gravity
- Maximum water holding capacity
- Percent pore space
- Volume expansion of 100 mL of soil

## Calculations

Let

Weight of Keen box plus filter paper = $a$ g

Weight of box plus filter paper plus air dry soil = $b$ g

Weight of box with wet saturated soil = $c$ g

Weight of box with wet residual soil, after removal of the wet expanded soil = $d$ g

Weight of box and the residual wet soil after drying at 105°C = $e$ g

Weight of watch glass = $f$ g

Weight of watch glass plus wet expanded soil = $g$ g

Weight of watch glass after drying at 105°C = $h$ g

% of moisture in air dry soil = $z$ g

Internal volume of the box = $v$ ml

- Apparent density = $(b - a)/v$
- Absolute specific gravity = $\dfrac{e-a}{v-(d-e)}$
- Maximum water holding capacity (%) = $\left(\left\{\dfrac{(c-a)-(b-a)}{(b-a)}\right\}\times 100\right)$
- Percentage of pore space = $\left(\dfrac{(d-a)-(c-a)}{v}\times 100\right) = \left(\dfrac{(d-e)}{v}\right)\times 100$
- Volume expansion of 100 mL soil (%) = $\left\{\dfrac{(g-h)+\left(\dfrac{h-f}{\text{sp.gr}}\right)}{v}\right\}\times 100$

***Note:*** Measurement of inner radius of Keen box is done by slide callipers. The vernier constant, is evaluated as follows:

Say 10 Vernier scale = 9 main scale division

1 Vernier scale = 9/10 main scale division

Vernier constant (V.C) = (1 – 9/10) mm = 0.1 mm = 0.01 cm

To determine the diameter record readings in the form of table shown below.

| **Main scale reading (cm)** | **Vernier scale division (V.D)** | **Vernier reading (V.D × V.C) cm** | **Total reading = diameter(d)** |
|---|---|---|---|
| No. of obs. Readings | Mean | No.of obs. Readings | Mean |
| 1 | | 1 | |
| 2 | | 2 | |
| 3 | | 3 | |

Now radius $r = \text{diameter } (d)/2$ cm.

To measure the height of the Keen box, tabulate as follows

| **Main scale reading (cm)** | **Vernier scale division (V.D)** | **Vernier reading (V.D × V.C) cm** | **Total reading = height = h cm** |
|---|---|---|---|
| No. of obs. Readings | Mean | No. of obs. Readings | Mean |
| 1 | | 1 | |
| 2 | | 2 | |
| 3 | | 3 | |

$$\text{Inner volume}(v) = \pi r^2 h$$

## 2.10 SOIL WATER CONTENT

### 2.10.1 Soil Moisture Per Cent (Direct Method)

Soil water content is generally reported as the ratio of the mass of water present in a soil sample to the mass of the sample after it has been dried to a constant weight. It is usually a dimensionless ratio which when multiplied by 100 gives the percentage value on a mass basis.

#### Procedure (Gravimetric method)

- Take 20 gm air dry soil sample in a weighed moisture can (with close fitting lid) and weigh again.
- Keep the open moisture can with soil in a hot air oven at 105 °C for 24 hrs.
- Close the can, transfer to a desiccator and let cool weigh.

#### Calculations

$$\text{Air dry moisture (\%)} = \frac{W_1 - W_2 \times 100}{W_2}$$

where $W_1$ = weight of air dry soil,

$W_2$ = weight of oven dry soil.

### 2.10.2 Neutron Probe Method (Indirect Method)

This is a method for determination of soil moisture status *in situ* without disturbing the system. In the neutron probe method number of hydrogen neuclei present per unit volume of soil is measured. Since hydrogen neuclei have a marked property for scattering and slowing neutrons, the same is exploited in the neutron method for measuring soil water content.

## Principle

Fast moving neutrons emitted from a radioactive source (usually Radium – Berrylium or Americium – Beryllium) upon collision with a particle having mass nearly equal to its own, like hydrogen atom in the soil, release their energy and gets thermalized or slowed down. The thermalized neutrons are detected by a detector and recorded on a scalar. Usually $BF_3$ gas is used as detector of slowed down neutrons. Increased thermalization indicates higher water content of the soil. The zone of influence is normally about 15–20 cm around the detector.

## Apparatus

- Neutron probe assembly consisting of probe, detector, scalar (counting device) and cable (Fig. 2.7).
- Access tube of aluminium or steel of 20 gauze with 1.9 inch and 2.0 inch internal and outer diameter, respectively.
- Soil auger slightly smaller than the tube for drilling the access holes.

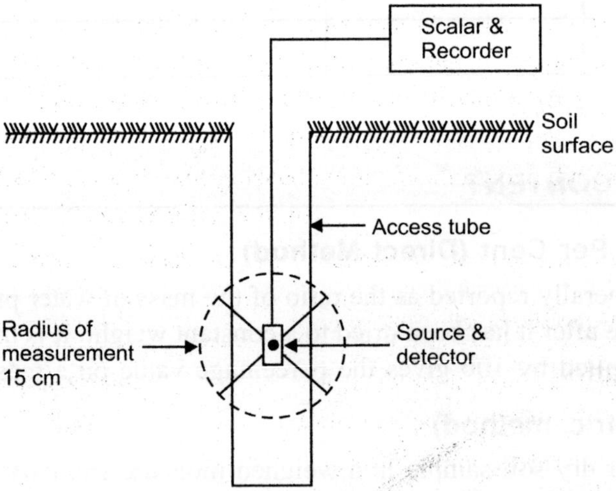


**Fig. 2.7** Neutron moisture meter for measuring soil water content

## Procedure

### Calibration

- Prepare a plot measuring 1 m × 1 m in the field.
- Drill a hole with the help of auger and insert the access tube in the soil with little disturbance such that no bulge is created in the access tube. Keep the access tube 10–20 cm above the soil and cover with inverted can or close its opening with a rubber cork to prevent entry of trash. In order to prevent water entry into the tube, close the lower end of the access tube with rubber stopper.
- Turn on the scalar and allow it to warm up for few minutes.

- Place the probe on the top of the access tube and measure the counts, called standard counts. The normal counting time is one minute. The 'background' count thus obtained should not be much more than 100 counts per minute. Approximately a 15 cm soil layer is characterized by a single measurement.
- Take readings at successive depth intervals starting at least 18–25 cm from the soil surface.
- Lower the probe in the access tube to a depth at which water content is to be determined and note the counts.
- Calculate the count ratio by dividing the observed counts at a depth by the standard counts.
- Determine the water content of that layer of soil gravimetrically and convert to volumetric water content by multiplying it with bulk density of the soil.
- Construct a calibration curve (Fig. 2.8) by filling a linear relation ($\theta_v = a + b$ CR) between volume water content ($\theta_v$) and the count ratio (CR).

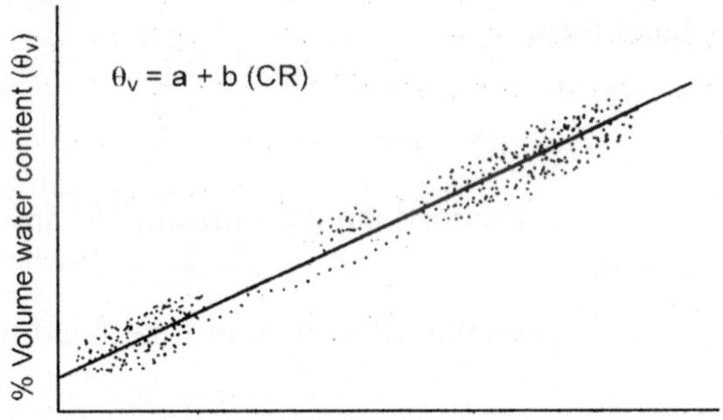


**Fig. 2.8** A schematic representation of a typical calibration curve for measuring soil water content by neutron probe method

## Moisture determination

- Install the access tube in the soil.
- Measure the standard counts by placing the probe on the top end of the access tube.
- Lower the probe into the access tube to the desired depth and note the counts.

## Calculations

$$\text{Standard count} = \text{SC}$$

$$\text{Actual counts} = \text{AC}$$

$$\text{Count ratio CR} = \frac{\text{AC}}{\text{SC}}$$

$$\text{Constants of calibration curve} = a \text{ and } b$$

$$\text{Volumetric water content, } \theta_v = (a + b\ \text{CR}).$$

## Precautions

- Use dent-free access tubes

- Always plug the lower end of the access tube
- Protect the neutron source from free water, otherwise it will get spoiled
- Do not touch open probe with hands
- Check the batteries of the probe and scalar before taking the instrument to the field.

## 2.11 DETERMINATION OF SATURATED HYDRAULIC CONDUCTIVITY IN LABORATORY

### 2.11.1 Constant Head Permeameter Method (For Very Porous Soils)

#### Principle

If a constant water head is maintained on one end of a saturated column of soil of length(L), the volume of water (Q) percolating through the other end per unit cross-sectional area (A) of the soil column per unit time (*t*) will be directly proportional to the hydraulic gradient ($\frac{\Delta H}{L}$) across the length of soil column. Therefore,

$$\frac{Q}{A \cdot t} = -K\frac{\Delta H}{L} \qquad ...(2.11.1.1)$$

According to Darcy's law the proportionality constant K in the above equation is the hydraulic conductivity of the soil.

The symbol $\Delta H = H_i - H_o$; denotes the difference in total head between inflow and outflow ends of a column.

Again
$$H_i = H_{si} + H_{gi} \qquad ...(2.11.1.2)$$
$$H_o = H_{so} + H_{go} \qquad ...(2.11.1.3)$$

Where $H_s$ and $H_g$ stand for suction head and gravitational head, respectively; *i* and *o* indicate inflow and outflow ends, respectively. Upward direction is considered as positive, water drips out freely from the bottom of the soil column. Under such conditions $H_o = 0$, since $H_{so}$ and $H_{go}$ both are equal to zero.

Thus
$$\Delta H = H_{si} + H_{gi} \qquad ...(2.11.1.4)$$
but
$$H_{gi} = L \qquad ...(2.11.1.5)$$
Therefore
$$\Delta H = H_{si} + L \qquad ...(2.11.1.6)$$
Hence
$$\frac{Q}{A \cdot t} = -K\frac{(H_{si} + L)}{L} \qquad ...(2.11.1.7)$$

#### Apparatus

- Brass permeameters of about 7 cm inside diameter and 10 cm length with perforated bottoms.
- A wooden or iron stand for supporting the permeameter

- Measuring cylinders
- Glass rods
- Stop watch
- A water reservoir with Mariotte arrangement to maintain a constant water head on the soil surface

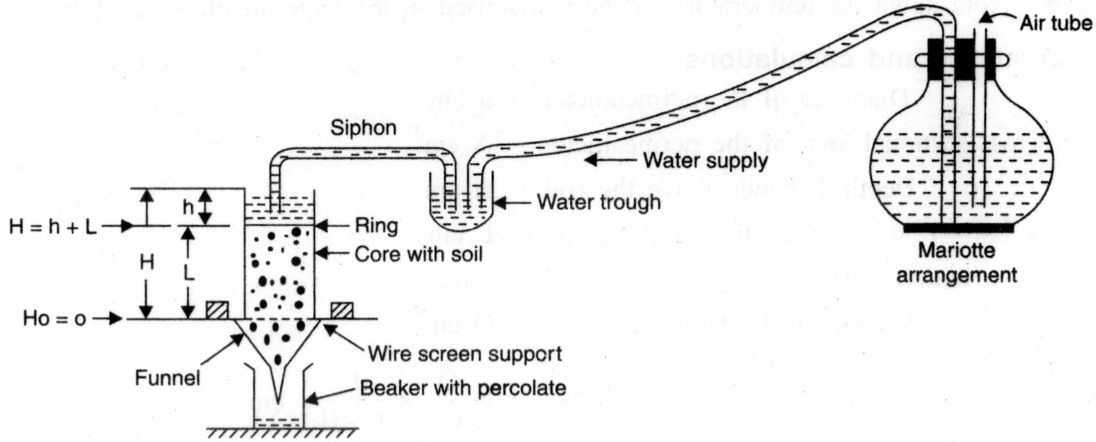


**Fig. 2.9** Apparatus for measuring saturated hydraulic conductivity with constant water head method

## Procedure

- Place a filter paper on the screen of the permeameter.
- Take 200 g of air-dry soil passed through a 2 mm sieve and dump the entire sample in one lot into the permeameter.
- Mix and pack the sample by tapping the permeameter 15–20 times on a wooden block through a height of 2.5 cm.
- Place a filter paper on the soil surface for protection against damage by washing when the water enters initially.
- Saturate the soil by placing the permeameter in a tray filled with water in such a way that the water level is slightly above the bottom of the samples.
- Leave it as such for overnight or longer till it is fully wet at the surface. The saturation point is indicated by a continuous and shinning water film at the soil top.
- Place the permeameter on the stand and start the siphon to ensure a constant head of 2–3 cm of water on the top of the soil by siphon tubes and Mariotte arrangement.
- Carry out at least 4 replicates to have an idea of measurement of variability.
- Record the time as soon as the water head on the soil top becomes constant and a steady flow is attained at the outflow end.
- When steady flow is reached start collecting the discharge in a measuring cylinder.
- Measure the volume of percolate collected in a known time.

- Record a few consecutive readings until the flux is constant.
- Measure the exact water head on the soil surface with the help of a meter ruler and then dismantle the experiment.
- Measure the length of soil column by pushing a glass rod vertically and note down the length of rod marked with soils.
- Note down the temperature of the water used in the experiment.

**Observations and calculations**

Diameter of the permeameter = $d$ cm

Cross-sectional area of the permeameter = A $cm^2$

Depth of water above the soil = H cm

Length of soil column = L cm

Time for which discharge collected = $t$ min

Volume of discharge collected = Q $cm^3$

$$\text{Hydraulic conductivity } K_s = \left(\frac{Q}{A \cdot t} \times \frac{L}{L+H}\right) \text{ cm min}^{-1}$$

### 2.11.2 Falling Head Method (For Slowly Permeable Soils)

**Principle**

In this method, drop in water level in a narrow tube is measured instead of flow. Let time taken by water to fall from initial head $H_1$ to final head $H_2$ be $t$ and let H be the head at any intermediate time. Now if '$-dH$' be the change in head in time interval $dt$ and $a$ is the cross-sectional area of the stand pipe, the rate of flow using Darcy's law is given by

$$Q = \frac{-dH \cdot a}{dt} = KA\frac{\Delta H}{L} \quad \text{...(2.11.2.1)}$$

where ($\frac{\Delta H}{L}$) is the hydraulic gradient, L is the length of soil column and A is the cross-section of the soil in the permeameter.

or $$KA\frac{\Delta H}{L} = \frac{-dH}{dt}a \quad \text{...(2.11.2.2)}$$

or $$\frac{(KA)}{a \cdot L}dt = \frac{-dH}{H} \quad \text{...(2.11.2.3)}$$

On integration, $$\frac{K \cdot A}{a \cdot L}\int_0^t dt = -\int_{H_1}^{H_2}\frac{dH}{H} \quad \text{...(2.11.2.4)}$$

or $$\frac{K \cdot A}{a \cdot L} \cdot t = \ln\frac{H_1}{H_2} \quad \text{...(2.11.2.5)}$$

or

$$K = \left[2.303\frac{(a \cdot L)}{A \cdot t}\log_{10}\frac{H_1}{H_2}\right] \qquad ...(2.11.2.6)$$

## Apparatus

Galvanized iron cylinder (40 cm in length, 30 cm diameter) with a conical top

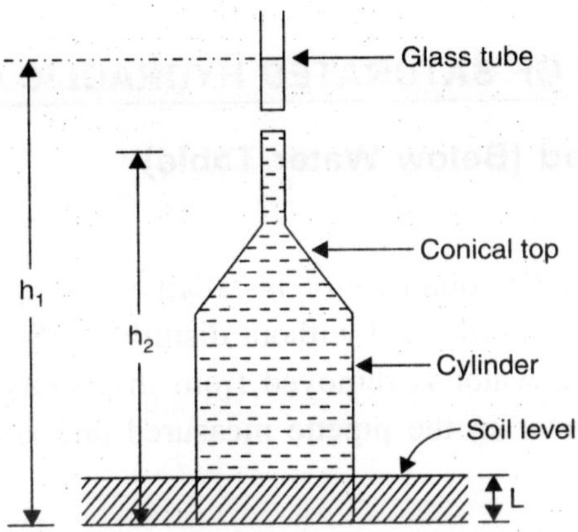


**Fig. 2.10** Falling head permeameter

## Procedure

- Press the cylinder into the soil to a known depth for which determination is to be made.
- Transfer the sample to the laboratory and fit the apparatus as shown in (Fig. 2.10).
- Wet the sample from below by supplying water to the bottom by means of a three way stop cock.
- Fill the space above the sample with water either by upward flow through the sample or by introducing water by a pipette at the top of the sample.
- Maintain a water level in the stand pipe somewhat above the level by introducing water through a three way cock.
- Connect the stand pipe to the sample by opening the stop cock and measure the time for water level to fall from $H_1$ to $H_2$.
- Repeat the steps and make additional measurements.

## Observations and calculations

Diameter of the stand pipe = $d$ cm

Area of the stand pipe = $a$ cm$^2$ = $\frac{\pi d^2}{4}$

Length of the sample = L cm

Diameter of the sample = D cm

Cross-sectional area of the sample = A cm$^2$ = $\frac{\pi D^2}{4}$

Initial hydraulic head = $H_1$ cm

Final hydraulic head = $H_2$ cm

Time taken for change in head = $t$ sec

$$\text{Hydraulic conductivity K} = \left[\frac{aL}{At}\ln\frac{H_1}{H_2}\right] \text{cm s}^{-1}$$

## 2.12 DETERMINATION OF SATURATED HYDRAULIC CONDUCTIVITY IN FIELD

### 2.12.1 Piezometer Method (Below Water Table)

#### Principle

For measuring saturated hydraulic conductivity, installation of piezometer tubes into an auger hole as big as the tube's diameter is performed without disturbing the soil. A cavity is then provided at the bottom of the pipe and water is removed from the cavity after elimination of puddling effect. The rate of rise of water in the pipette measured and conductivity is calculated.

#### Apparatus

- Piezometer tube: Aluminium pipe of diameter of 5 cm (Fig. 2.11)
- Auger; screw type that fits inside the piezometer tube.
- Water pump or bailer
- A device to measure the depth of water in the cavity
- Stop watch

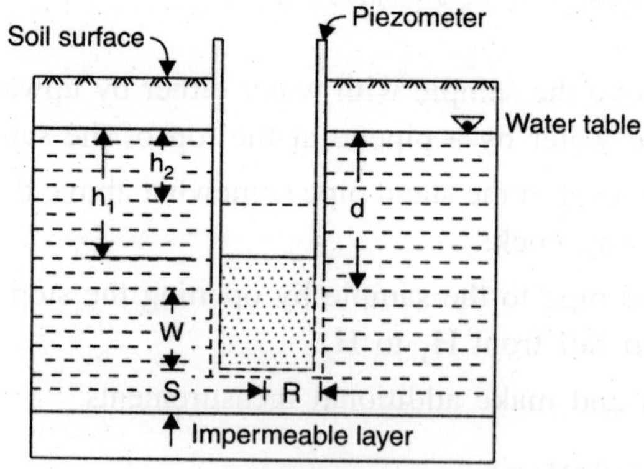


**Fig. 2.11**

#### Procedure

- Remove plant material, rubbish waste and loose soil from the area.
- Bore a hole up to a depth of 10 cm, remove the auger and insert piezometer pipe into the hole. Dig out an additional 10–15 cm soil by inserting auger into the pipe and tap the pipe into the excavated hole, thereby lowering the pipe to a desired depth.

- Eliminate puddling effect by inserting a tube down the pipe into the cavity and remove water with the help of a bailer.
- Allow the water to rise in the pipe and remove the water from the cavity. Repeat until constant rate of water rise is attained.
- Record the difference between the depths of water table and of water level in the pipe at two times $t_1$ and $t_2$.

### Calculations

Radius of piezometer tube $= R_p$ meter

Difference between depth of the piezometer tube and the water tube $= d$ meter

Shape factor $=$ S meter

where

$$S = \frac{R_p \cdot d}{0.15}$$

Difference in time $= \Delta t$ sec

$$\text{Hydraulic conductivity} = \frac{\pi R_p^2}{2S\Delta t} \ln \frac{h_1}{h_2}$$

## 2.12.2 Inverted Auger Hole Method (Above Water Table)

### Principle

A hole is bored to the desired depth and a constant head of water is maintained in the hole. The fall of water level in the whole is measured under steady state condition.

### Apparatus

- Spade
- Bailer with pulley
- Measuring tape
- Stop watch

### Procedure

- Bore an auger hole in the soil to a given depth.
- Measure the depth and diameter of the auger hole.
- Fill the auger hole with water.
- Measure the initial depth of water inside the hole.
- Measure the periodic fall of water level inside the hole.
- Determine the hydraulic conductivity as shown below.

### Calculations

Diameter of the auger hole = D meter

The initial depth of water inside the hole at time $t_0$ (min) = $h_o$ meter

The depth of water inside the auger hole at time $t_1$ (mm) = $h_1$ meter

Determine the slope of the curve by plotting log $(h_1 + \frac{D}{2})$ vs $t_i$. This is the value of tan θ or periodic depth of water inside the hole $h_i$ at time $t_i$.

$$\text{Also } \tan\theta = \frac{\log\left(h_o + \frac{D}{2}\right)}{t_1} - \frac{\log\left(h_1 + \frac{D}{2}\right)}{t_0}$$

$$\text{Hydraulic conductivity} = 1.15\ \text{D} \tan\theta\ \text{m min}^{-1}$$

## 2.13 INFILTRATION

### Principle: Cylinder Infiltrometers

Such type of instruments are used most commonly. Usually, metal rings of known diameter are driven into the soil to depth ranging from a few inches to more than a foot, so that lateral or divergent flow of water from the rings may be reduced to a minimum. The method of water addition to the cylinders includes such principles as constant heads, falling heads etc. Previously only single rings were used to study infiltration rates but recently double or multiple ring devices are employed in order to check the lateral movement of water to a still higher degree. In such device two or more rings are pushed into the soil surrounding each other, isodiametrically. Measurements of infiltration rates in the central compartment are thought to be indicative of the vertical component of flow.

### Equipments

- Cylinders made of 14 to 16 gauze iron sheet, rolled in circular fashion and joints ground to a smooth finish. One end of the cylinder is sharpened from outside keeping the inside completely smooth. This facilitates easy penetration of the cylinders inside the soil. The inner diameter of the central ring ranges between 30 and 35 cm and that of outer ring between 40 and 45 cm. The weight of the rings should be between 40 and 45 cm.
- Circular driving caps to fit over each of the rings.
- Hammer of enough weight to push the rings into the soil.
- Watch
- Hook gauge

### Procedure

- Drive the first central ring vertically downwards into the soil at a suitable spot in the field to a depth of 15–20 cm by hammering on the central guide rod of the circular cap in such a way that the ring penetrates the soil straight downwards from all sides.

- Tap soil into the space between the soil-column and the cylinder to bring the soil inside the ring to its natural condition as far as practicable.
- Drive the outer ring into the soil iso-diametrically with the central ring.
- Apply 10–15 cm water inside the central ring and also in the space between the two rings.
- Place the hook gauge in the central ring.
- Record the receding water level against time at suitable time intervals in the central ring.
- Express the rate of infiltration using values averaged over time intervals in cm $hr^{-1}$ or inches $hr^{-1}$.

## 2.14 SOIL MOISTURE CONSTANTS

### 2.14.1 Hygroscopic Coefficient

The amount of moisture taken up by a dry soil when kept in contact with an atmosphere saturated with water vapour (100% relative humidity) at a given temperature is known as the hygroscopic coefficient of the soil at that temperature; and is usually expressed on the dry weight basis.

#### Equipments

- Augers for soil sampling
- Moisture cans
- Desiccator
- Watch glasses
- Drying oven
- Balance
- 3.3% $H_2SO_4$ to give a relative humidity of 98% at a vapour tension of 31 atmosphere.
- 2 mm sieve

#### Procedure

- Take 5 g duplicate samples of air dry soil passed through 2 mm sieve on watch glasses in a desiccator containing 3.3% $H_2SO_4$.
- Allow to equilibrate for 7 days.
- Rapidly transfer to tared moisture cans and weigh accurately.
- Calculate the moisture percentage at hygroscopic coefficient after oven drying at 105 °C for 24 hours.

[*Note:* Higher the clay content, higher will be the hygroscopic coefficient. Similarly organic matter also increases the hygroscopic coefficient. Hygroscopic water which is practically unavailable to plant forms large proportions of soil water in heavy textured soils].

### 2.14.2 Moisture Equivalent

It is usually defined as the amount of moisture held by a soil 1 cm thick when subjected to a centrifugal force of 1000 times that of gravity for 30 minutes corresponding to 2400 r.p.m of the centrifuge.

### Equipments

- Moisture cans
- Oven
- Balance
- Spatula
- Water trough
- Whatman No. 2 filter paper
- 2 mm sieve
- Sampling auger
- Centrifuge (Briggs and Mcclane centrifuge with sample cups).

### Procedure

- Cut Whatman No. 2 filter paper (square shape) and fix properly on the wire gauged bottom of the moisture equivalent boxes.
- Weigh accurately 30 g of air dry sample which has been passed through 2 mm sieve and place in the box.
- Tap and level the surface of the soil.
- Place the boxes in 1 cm water and allow to saturate for 24 hr. After saturation, take them out of water, wipe out the moisture, put their covers and transfer to the centrifuge arranging them in opposite direction for proper balancing.
- Centrifuge at 2400 r.p.m. for 30 minutes. [Bring the centrifuge up to the required speed in 3 minutes. After 30 minutes stop the machine and bring to rest within 3 minutes].
- Transfer the soil to weighed moisture boxes and weigh.
- Oven dry the samples and weigh again.
- Calculate the percentage of moisture at moisture equivalent.

### Calculations

$$\text{Moisture equivalent (M.E)\%} = \frac{b-c}{c-a} \times 100$$

where $a$ = weight of empty box with lid and filter paper,

$b$ = weight of box plus soil at its M.E point,

$c$ = final dry weight of box plus dry soil.

### 2.14.3 Field Capacity

Field capacity is defined as the amount of water held in soil after excess gravitational water has drained away and after the rate of downward movement of water has materially decreased (Veihmeyer and Hendrickson, 1931). Such a situation is normally reached 48 to 72 hours after saturation. Sandy soils reach field capacity earlier than clayey soil. The field capacity is the upper limit of available soil moisture range in soil-water-plant relationship. The force with which moisture is held in the soil ranges between 0.1 and 0.33 bars.

### 2.14.3.1 Laboratory method

#### Equipments

- A glass cylinder (45 cm in length and 6–7 cm in diameter)
- Watch glass to cover open end of cylinder
- 2 mm sieve
- Balance
- Drying oven
- Soil auger
- Moisture cans
- Parraffin wax
- Deaerated water

#### Procedure

- Pack slowly the soil where field capacity is to be determined (after passing through a 2 mm sieve) uniformly in the clean and dry glass cylinder, leaving about 10 cm of the top of the cylinder unfilled with soil.
- Take care that no air pocket is created inside the glass cylinder.
- To maintain the similar compactness as in the field, determine the bulk density of the soil.
- Then fill the cylinder in such a way that the same compactness as in field condition is maintained. This may be done by measuring the volume of the cylinder and the amount of oven dry soil which is field in this volume to bring the soil to the required bulk density level.
- After completion of soil packing in the cylinder, a glass tube of small diameter is pushed inside the soil at the centre taking care that the capillary of the glass tube is not filled with soil so that the passage of the displaced air is facilitated.
- Then apply sufficient amount of de-aerated water onto the soil surface of the glass cylinder so as to saturate completely the top about 25 cm of the soil leaving about 10 cm of the bottom soil dry. Seal the upper surface of the soil with paraffin wax and cover with a watch glass to prevent evaporation loss from the surface.
- Plug the open end of the glass tube with cotton wool to minimise evaporation losses.
- Make the determinations in duplicate.
- Allow to stand for 48–72 hours and when excess water has drained below, take samples from the wet zone leaving about 10 cm of the top soil.
- Determine the moisture content at field capacity after drying the soil sample in the oven.

***Note:*** The laboratory method is rapid and is commonly used when the water table is shallow and field method cannot be used. Value of field capacity by laboratory method usually does not coincide with that of field method since in the laboratory the natural conditions of the soil are disturbed. Normally field method is recommended for determining field capacity of the soil.

## Calculations

$$\text{Weight of moisture box} = m_b \text{ gram}$$

$$\text{Weight of moisture box + wet soil} = m_{bws} \text{ gram}$$

$$\text{Weight of moisture box + over dry soil} = m_{bds} \text{ gram}$$

$$\text{Percent water content} = \frac{m_{bws} - m_{bds}}{m_{bds} - m_b}$$

### 2.14.3.2 Field method

## Equipments and materials

- Black polythene sheet or straw mulch
- Moisture cans
- Spade and Auger
- Balance
- Drying oven
- Water

## Procedure

- Select a uniform plot of 3 m × 3 m. Remove weed, pebbles etc. and bund from all sides.
- Fill the plot with sufficient water to completely saturate the soil to the desired depth (Water table should not be within 2 m from the layer of which field capacity is to be determined).
- Cover the area with a thick straw mulch or polythene sheet to prevent evaporation loss.
- Take soil sample from centre of the plot from the desired layer and determine the soil moisture content daily until the value of two successive days are nearly equal.
- Plot the readings on a graph paper. The lowest reading may be taken to represent the value of field capacity of the soil.

### 2.14.4 Permanent Wilting Point

The permanent wilting point is that soil water content at which plants are unable to absorb water and wilt permanently. A plant is said to be permanently wilted when it will not regain its turgidity even after being placed in a saturated environment. Permanent wilting percentage is often characterized as the lower limit of available soil moisture.

Inspite of the fact that wilting point is a good indicator of lower limit of available water, there is enough evidence to indicate that wilting point is not a true intrinsic soil property. As such there does not exist an unique soil water retentivity value at which the water uptake by plants suddenly ceases, rather plant usually wilts at a point controlled by rate factor (both supply and demand). However, the 15 bar percentage has been found to be closely correlated with the permanent wilting point (Richard and Weaver 1943).

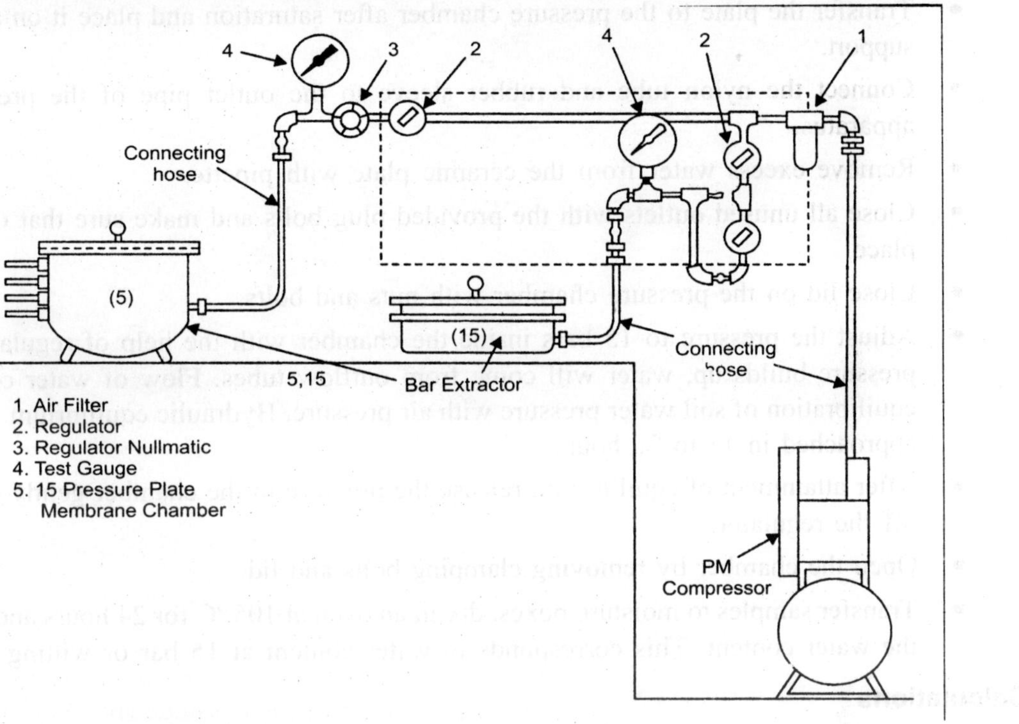


**Fig. 2.12** Pressure plate apparatus

## Equipments

- Pressure-membrane apparatus complete with all fittings and 15 bar ceramic plate.
- Brass soil retaining rings (1 cm high and about 6 cm in diameter which can held about 25 g soils).
- Drying oven
- Balance
- Moisture cans
- Syringe or pipette
- 2 mm sieve
- Soil sampling auger

## Procedure

- Take air dry soil samples passed through 2 mm sieve.
- Place the soil samples in rings in duplicate in the the ceramic plate.
- Level the samples in the ring.
- Saturate the samples by placing the plate with the rings in a trough of water for at least 24 hours. Water must be just enough to reach the upper edge of the rings.

- Transfer the plate to the pressure chamber after saturation and place it on a triangular support.
- Connect the nylon tube and rubber sleeve to the outlet pipe of the pressure plate apparatus.
- Remove excess water from the ceramic plate with pipette.
- Close all unused outlets with the provided plug bolts and make sure that *O* ring is in place.
- Close lid on the pressure chamber with nuts and bolts.
- Adjust the pressure to 15 bars inside the chamber with the help of regulator. As the pressure builds up, water will come from outflow tubes. Flow of water ceases upon equilibration of soil water pressure with air pressure. Hydraulic equilibrium is normally approached in 18 to 20 hours.
- After attainment of equilibrium, release the pressure in the chamber gently by shutting off the regulator.
- Open the chamber by removing clamping bolts and lid.
- Transfer samples to moisture boxes, dry in an oven at 105 °C for 24 hours and determine the water content. This corresponds to water content at 15 bar or wilting point.

## Calculations

Weight of the moisture box = $W_b$ gram

Weight of the moisture box + wet soil = $W_{bws}$ gram

Weight of the moisture box + over dry soil = $W_{bds}$ gram

$$\text{Per cent water content} = \frac{W_{bws} - W_{bds}}{W_{bds} - W_b} \times 100$$

### 2.14.5 Moisture Retention Curve

Similar to 15-bar moisture content, moisture contents at different values of suction like 0.1, 0.3, 1.0, 5.0 and 10 bars can be determined. The curve showing relationship between soil moisture content and suction is known as moisture retention curve.

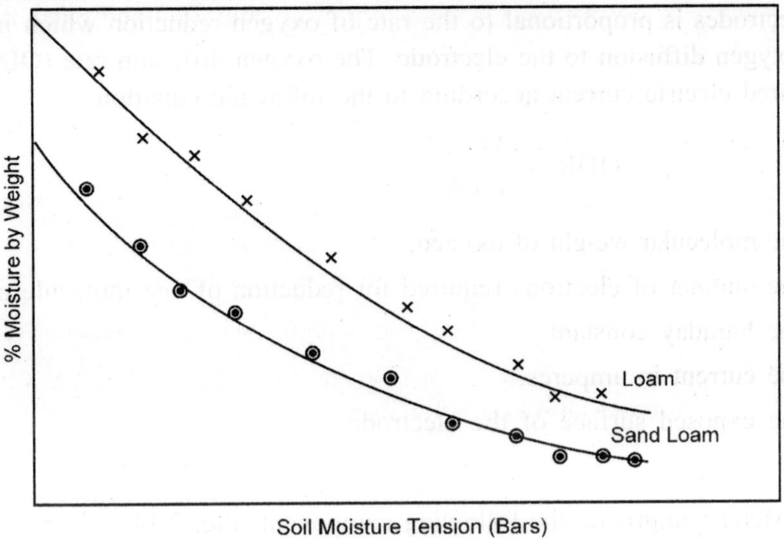


**Fig. 2.13** A schematic representation of a typical moisture release curve

### 2.14.6 Available Water

The available water in a soil is the amount of water which can be utilized by the plants for their normal growth and development.

Available soil moisture (%) = moisture at field capacity (%) – moisture at wilting point(%)

If bulk density is known, percent values can be converted into cm of water a foilows:

$$\text{cm of water} = w \times D_b \times \frac{d}{\rho} \times 100$$

where $w$ = percent of water on dry weight basis,

$D_b$ = bulk density of soil in Mg $m^{-3}$,

$\delta$ = density of water in Mg $m^{-3}$,

$d$ = depth of soil layer in cm.

## 2.15 OXYGEN DIFFUSION RATE (ODR)

For the purpose of ascertaining the movement of oxygen from atmosphere to actively respiring cells of plant root in the soil system determination of oxygen level at the interface between the root surface and soil is extremely important. Since the active root surfaces are covered with water films, air moves through air-liquid boundary. Movement of air occurs by diffusion in the water film-cell wall portion of the oxygen path. Lemon and Erickson (1952) proposed a method for measuring oxygen diffusion in soil with the help of platinum micro electrode.

### Principle

Upon application of a certain potential across the platinum electrode and a reference electrode inserted in the soil, oxygen is reduced at the platinum surface. The electric current flowing between the electrodes is proportional to the rate of oxygen reduction which in turn is related to the rate of oxygen diffusion to the electrode. The oxygen diffusion rate (ODR) is calculated from the measured electric current according to the following equation.

$$\text{ODR} = \frac{\text{M} \times i}{n\,\text{FA}} \quad \text{...(2.15.1)}$$

where M = the molecular weight of oxygen,

$n$ = the number of electrons required for reduction of one molecule of oxygen,

F = the Faraday constant,

$i$ = the current in amperes,

A = the exposed surface of the electrode.

## Apparatus

- ODR Meter comprising the following component (Fig. 2.14)
  Platinum electrode (cathode)
  Reference electrode (saturated calomel anode)
  Electric circuit to apply an electric potential of 0.65V.
  A milli-ampere meter to measure the output current.

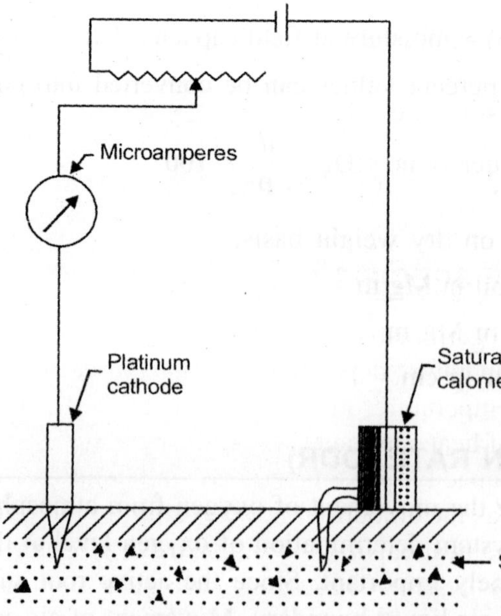


**Fig. 2.14** Schematic diagram of oxygen diffusion rate (ODR) measurement

## Procedure

### Preparation of platinum cathode

- Cut the copper wire to the required length and remove the insulation at both ends

- Solder a piece of platinum wire 8–10 mm long (22 gauge) to the copper wire and insulate the junction.
- Mount a sheath of plastic material leaving 4–5 mm platinum wire exposed.

### ODR measurement

- Insert the platinum electrode into the soil for measurements at shallow depths and ensure that there is good contact between the reference electrode and the soil.
- Apply 0.65 volts potential across the electrodes
- Wait for 5 minutes to attain steady state current
- Measure the output current with a micro-ampere meter between the reference cell and platinum electrode; calculate ODR.

### Calculations

Output current = $i$ μ ampere

Electrode length = $l$ cm

Electrode radius = $r$ cm

Surface area of electrode = A cm$^2$ = $(2\pi rl + \pi r^2)$ cm$^2$

Molecular weight of oxygen = M gram = 32 gram

Faraday constant = F coulombs/mole of $O_2$ = 96500

Number of electrons required for reduction of one molecule of oxygen = $n$ = 4

$$\text{Oxygen diffusion rate (ODR)} = \left(\frac{M \times i}{n\,FA}\right) \ \mu g\ cm^2\ s^{-1}$$

[*Note:* ODR value of $20 \times 10^{-8}$ g cm$^{-2}$ s$^{-1}$ or more suggests sufficient oxygen supply for root growth.]

## 2.16 DETERMINATION OF SPECIFIC HEAT OF SOIL

Soil temperature influences seed germination, root and shoot growth in soil. The rates of chemical and biological reactions are temperature dependent and also the rate of crop growth is influenced by the soil temperature. The temperature, in turn, depends directly upon the heat capacity and specific heat of soil. The ratio of heat supplied to its corresponding rise in temperature is the heat capacity of the soil. The heat capacity, per unit mass of soil solid is known as the specific heat of soil. Specific heat is defined as the amount of heat in calories required to raise the temperature of 1 g of soil solids by 1°C. The heat capacity is an extensive property of the system and varies with amount of material in the system whereas specific heat is an intensive property and is independent of the size of the system. The specific heat of soil varies from mineral to mineral. For most mineral soils specific heat is about 0.2 cal g$^{-1}$ °C$^{-1}$ whereas that for dry soil varies between 0.17 and 0.29 cal g$^{-1}$ °C$^{-1}$ . The knowledge of specific heat is extremely useful for evaluating the process of heat transfer through the soil. Also determination of specific heat at constant pressure allows are to calculate themodynamic quantities like enthalpy, entropy, free energy change of the system.

## Principle

When two bodies at different temperature are brought in close contact with each other, the heat is lost by the body at a higher temperature and gained by the one at lower temperature, ultimately resulting in attainment of thermal equilibrium when the two bodies are said to have the same temperature. This phenomenon occurs in an isolated system where no heat enters or leaves the system and no chemical interactions takes place between the bodies kept in contact. Thus, the specific heat of one of the bodies can be determined if that of the other is known.

Let an oven dry soil having mass ($m_1$), specific heat ($s_1$) and initial temperature ($\theta_1$) be dropped all of a sudden into water contained in a calorimeter with mass, specific heat and initial temperature as $m_2$, $s_2$ and $\theta_2$ respectively. If $m_3$ and $s_3$ are the mass and specific heat of the calorimeter and if the resulting temperature of soil-water mixture at equilibrium is $\theta$ then according to the principle of calorimetry (Heat gained = Heat lost); it follows that

$$\text{Heat gained by soil} = \text{heat lost by water and calorimeter} \quad ...(2.16.1)$$

$$m_1\, s_1\, (\theta - \theta_1) = m_2 s_2\, (\theta_2 - \theta) + m_3 s_3\, (\theta_2 - \theta) \quad ...(2.16.2)$$

or

$$m_1\, s_1\, (\theta - \theta_1) = (m_2 s_2 + m_3 s_3) \cdot (\theta_2 - \theta) \quad ...(2.16.3)$$

or

$$s_1 = \left[\frac{(m_2 s_2 + m_3 s_3) \cdot (\theta_2 - \theta)}{m_1 (\theta - \theta_1)}\right] \quad ...(2.16.4)$$

where $\theta_1 < \theta < \theta_2$.

## Apparatus

- Calorimeter with an insulation box
- Thermometer
- Weighing balance
- Tripod stand

## Procedure

- Weigh the calorimeter.
- Heat water to a temperature 20°C–25°C higher than the room temperature and pour about 50–60 mL of this to the calorimeter. Keep it in the insulation box and cover it.
- Weigh 25 g of soil
- Read the temperature ($\theta_1$) of the soil to the nearest 0.01°C.
- Record the temperature ($\theta_2$) of the water in the calorimeter to the nearest 0.01°C.
- Remove the lid and drop the soil immediately into the calorimeter and cover it again.
- Stir the suspension slowly until thermal equilibrium is reached.
- Record the final equilibrium temperature ($\theta$) of the calorimeter.

## Calculation

$$\text{Weight of soil} = m_1 \text{ gram} = 25 \text{ gram}$$

$$\text{Weight of calorimeter} = m_3 \text{ gram}$$

$$\text{Weight of calorimeter + hot water} = m_4 \text{ gram}$$

$$\text{Weight of hot water} = (m_4 - m_3) = m_2 \text{ gram}$$

$$\text{Temperature of soil} = \theta_1 \text{ °C}$$

$$\text{Temperature of hot water in the calorimeter} = \theta_2 \text{ °C}$$

$$\text{Equilibrium temperature} = \theta \text{ °C}$$

$$\text{Heat capacity of water} = 1 \text{ cal g}^{-1} \text{ °C}^{-1}$$

$$\text{Specific heat of copper (calorimeter)} = 0.093 \text{ cal g}^{-1} \text{ °C}^{-1}$$

$$\text{Specific heat of the soil } (s_1) = \left[\frac{(m_2 \times 1 + m_3 \times 0.093) \cdot (\theta_2 - \theta)}{25(\theta - \theta_1)}\right] \text{cal g}^{-1} \text{ °C}^{-1}$$

●●●

# 3 Soil Chemistry

## 3.1 ELECTROMETRIC MEASUREMENT OF SOIL pH

### Principle

The pH value which is a measure of the hydrogen (or hydroxyl) ion activity of the soil water system indicates whether the soil is acidic, neutral or alkaline in reaction. The pH value of the solution surrounding the soil particles fluctuates in the natural state because of changing soil: solution relationships brought about by climate, cultivation, crop growth and other factors. A sample of soil may have a certain pH value at the time it is sampled in the field which probably changes as the sample is dried and prepared for laboratory analysis. In the laboratory the sample is subjected to rewetting with water or certain salt solutions as the case may be, to establish the probable range of pH values it would have in its natural state. Since, crop growth suffers much under very low (strongly acidic) as well as high pH suitable reclamation measures become necessary. Presence of neutral soluble salts (as in the case of saline soils) is not normally reflected in the pH but when in large excess they tend to suppress the ionic activity and gives rise to what are known as 'activity errors'. These errors can also be significant when a small amount of dilute solution is being measured when potassium chloride from the salt bridge causes changes in activity. The salt effect is overcome or rather standardized, by taking pH measurements in potassium chloride solution rather than in water. Usually pH values in 1(N) KCl are lower than in water.

Soil acidity is known to exert adverse effect on crop growth by its effect on nutrient availability and microbial activity. Measurement of soil pH only is not a true representation of soil acidity. This is due to the fact that soil pH which is a measure of active acidity is subject to change due to a number of factors. Substantial amounts of soil acidity reside in the soil solid phase, in the interlayer spaces, as solid phase minerals and in the functional group of soil-organic fraction. All these contribute to the pool of active acidity in response to any shift in the thermodynamic equilibrium and thus helps in maintaining the buffering capacity of soils.

### Suspension effect in soil pH measurement

In the pH measurement the reference and indicator electrodes are immersed in a heterogeneous soil suspension comprising of dispersed solid particles in an aqueous solution. If the solid particles are allowed to settle down, the pH can be measured in the supernatant liquid or in the sediment. Placement of electrode pair in the supernatant normally gives a higher pH reading than placement of the same in the sediment. The difference in soil pH reading is called the **Suspension effect.**

In practice a measured quantity of soil is shaken with a convenient volume of water or salt solution under consistent conditions and the pH of the suspension is determined electronically on a direct-reading pH meter, using a glass electrode with a saturated KCl — calomel reference electrode. Usually soil : water ratio of 1 : 2.5 or 1 : 1 is used for routine analysis. (For more background theory and information see Chapter 1).

### Reagents and apparatus

- Standard buffer solutions

At least, two buffer solutions must be used one at low pH range and the other at high pH range. The buffer solution most commonly used for low pH values is a 0.05 M solution of potassium hydrogen pthalate which has a pH of 4.001 at 20°C. The pH of this buffer varies with temperature from 4.000 at 15°C to 4.020 at 35°C. The reagent must be of highest purity and the water used for its solutions should be double distilled water. The buffer can be stored in well sealed Pyrex or polythene bottles but preferably should be freshly prepared every 2 weeks. For high pH buffer solution a 0.01M solution of borax is convenient and at 20°C has a pH of 9.22. This buffer if protected from atmospheric carbondioxide can be kept for about a month. Three to four drops of toluene addition prevents growth of mould.

Standard certified buffer tablets are available for a whole range of pH-values and these are most convenient. One tablet dissolved in a specified amount of water, usually 100 cc provides a solution of known pH to within about 0.02 of a unit. Buffer tablets of pH 4.0, 7.0 and 9.2 are mostly used now a days for instrument calibration. For 1(N) KCl:Analytical Grade salt is used.

- Glass electrode pH meter with calomel reference electrode and salt bridge.
- 50 mL or 100 mL beakers, short stirring rods and distilled water wash bottle.

### Procedure

- Weigh 10 g soil sample in a 50 mL beaker and add 25 mL of distilled water (soil: water ratio of 1:2.5).
- Stir the suspension at regular intervals for 30 minutes.
- Measure the pH with the glass electrode stirring the suspension well just before immersing the electrode.
- Switch on the pH meter at least 15 minutes before for allowing it to warm up and standardize the glass electrode using standard buffers.
- Adjust the temperature compensation knob to the temperature of the test solution (More theoretical aspects are dealt in the Chapter 1).
- Rinse the electrode with distilled water after each determination and remove water from the surface with a piece of blotting paper.
- Check the standardization process after every ten determination.

- To determine pH in 1.0 (N) KCl, use 25 mL of 1.0 (N) KCl instead of water and record the pH after stirring intermittently for one hour.

***Notes:***

- pH determination of soil in water and KCl provides information on the nature of charge distribution on soil colloids, which may have a far-reaching effect on nutrient management and utilization.
- pH values should always be reported precisely, according to the ratio and liquid used.
- The glass electrodes must be dipped in water when not in use and it is to be ensured that the reference electrode always contains saturated potassium chloride solution in contact with solid potassium chloride crystals.
- $\Delta pH = pH_{H_2O} - pH_{KCl}$

  $\Delta pH > 0$ implies negatively charged clay surface

  $\Delta pH < 0$ implies positively charged clay surface

  $\Delta pH = 0$ implies zero point charge (ZPC), *i.e.*, the pH at which surface charge becomes zero.

## 3.2 DETERMINATION OF BUFFERING CAPACITY OF SOIL

### GENERAL DISCUSSION AND PRINCIPLE

#### Buffer action

Chemically a buffer solution is defined as one that resists a change in pH on addition of acid or alkali. Buffer solutions contain compounds that react with both acid or base so that the $H^+$ ion concentration in the solution remains constant. The buffer solutions thus have 'reserve acidity' or 'reserve alkalinity'. Buffer solutions usually consist of a mixtures of solutions of a weak acid or base and its salt, for example, acetic acid and sodium acetate. The pH of the buffer is governed by salt-acid ratio and the ionisation constant of the acid and is given by Henderson equation, pH = pKa + log $C_{salt}/C_{acid}$ (C = concentration). The resistance to the change in pH on addition of an acid or alkali is called buffer action. This buffer action is measured by 'buffer capacity' (β). It is the amount of a strong base required to produce unit change of pH of the solution.

$$\beta = \frac{db}{d(\text{pH})} \qquad ...(3.2.1)$$

where *db* amount of added base causes *d*(pH) change in pH. The buffer capacity (β) is maximum when the acid and salt are present in equal concentrations. Let *a* and *b* be the concentration of the acid and alkali mixed to produce a buffer, *a* > *b*. The concentration of the salt $C_s = b$, and final concentration of acid $C_a = a - b$. Applying **Henderson equation:**

$$\text{pH} = \text{pKa} + \log \frac{C_s}{C_a} = \text{pKa} + \frac{1}{2.303} \ln \frac{b}{a-b} \qquad \text{...(3.2.2)}$$

Therefore,

$$\frac{d(\text{pH})}{db} = \frac{1}{2.303} \frac{a-b}{b} \frac{b}{(a-b)^2} \qquad \text{...(3.2.3)}$$

or

$$\frac{db}{d(\text{pH})} = 2.303 \frac{b(a-b)}{a} \qquad \text{...(3.2.4)}$$

*i.e.,* Buffer capacity

$$\beta = 2.303 \frac{b(a-b)}{a} \qquad \text{...(3.2.5)}$$

or

$$\frac{d\beta}{db} = \frac{2.303(a-2b)}{a} \qquad \text{...(3.2.6)}$$

When $b = a/2, \quad d\beta/db = 0$

that is β is maximum or minimum at half neutralisation point.

But

$$\frac{d^2\beta}{db^2} = -\frac{2.303 \times 2}{a}, \text{ which is negative} \qquad \text{...(3.2.7)}$$

Hence, the half neutralization point is maximum for β.

The buffering capacity is best when $C_{acid} = C_{salt}$ in the buffer mixture *i.e.,* when pH = pKa. Thus, a sodium acetate–acetic acid buffer's capacity is maximum when equimolecular concentrations are taken in their mixture; the pH becomes 4.74 (for $Ka = 1.82 \times 10^{-5}$). When the ratio salt/acid is varied, the buffer will have a different pH, but not far away from the value of pKa. The maximum variation in the acid: salt ratio allowed is 1:10 or 10 : 1. The limiting values of the buffer pH becomes

$$\text{pH} = \text{pKa} \pm 1.$$

**Some of the common buffers used in the laboratory with their pH ranges:**

| Buffer solutions | pH range |
|---|---|
| • Phthalic acid and pot acid pthalate | 2.2–3.8 |
| • Acetic acid and Na–acetate | 3.7–5.57 |
| • Pot-acid pthalate and di-potassium pthalate | 4.0–6.2 |
| • $Na_2HPO_4$ and $NaH_2PO_4$ | 5.9–8.0 |
| • Boric acid and Borax | 6.8–9.2 |
| • Borax and $NaOH^-$ | 9.2–11.0 |

## Soil as a buffer system

In soils clay and humic fractions act as buffer systems. Certain soils such as heavy clays and peats have greater reserves of acidity than say, sands and such soils with large reserves of either acidity or alkalinity are said to be well buffered. The magnitude of reserve acidity usually far exceeds that of the active acidity. In clay soils high in organic matter reserve acidity was even 50,000 to 100,000 times greater than active acidity. The larger the buffer capacity the larger is the amount of lime required to raise the soil pH to desired level. Aluminium is partly responsible for the buffering action of soils because as the pH value rises aluminium dissociates hydrogen ions from coordinated water molecules in the clay.

One can visualize a situation when lime is added to an acid soil for soil pH amelioration. The $H^+$ ions in the soil solution would be neutralized; but fresh set of $H^+$ ions will come in the soil solution from the pool of adsorbed $H^+$ ions on the soil colloids. Consequently the resulting pH rise would be very small and would remain so until sufficient lime is added to exhaust appreciably the reserve acidity (adsorbed $H^+$ ions). The buffering capacity of a soil depends on its cation exchange capacity (CEC). Higher the CEC, greater would be the buffering capacity of the soil. This is due to the fact that the reserve acidity must be neutralized to effect a given rise or lowering of percentage of base saturation. Thus a fine textured acid soil with high cation exchange capacity requires more lime as compared to about the same change in pH.

### Reagent

Standard 0.01 (N) NaOH solution. (Standardized against a standard acid).

### Procedure

- Weigh 10 g soil in a 50 mL beaker.
- Add 25 mL distilled water and stir intermittently for half an hour.
- Measure the pH.
- Also measure the pH of 25 mL distilled water used for the experiment.
- Add 15 mL increments of 0.01 (N) NaOH (15, 30, 45, 60, 75, 90, 105, 120, 135, 150 mL) from a burette and read the pH of soil and of water.
- Plot the data with pH on the ordinate and volume of alkali (mL NaOH) added on the abscissa.

### Calculations

From the curve record the ml of NaOH required for the soil at inflection point *i.e.*, the threshold point at which the curve suddenly shoots up as all the acid is neutralized and thereafter the pH rise occurs due to concentration of the base added. Let volume of NaOH consumed at inflection point be $x$ mL.

meq of NaOH consumed in the titration = $\chi \times 0.01$

Now, 10 g of soil has a buffering capacity of $\chi \times 0.01$ meq.

So, 100 g soil has the buffering capacity of

$$\left[\chi \times 0.01 \times \left(\frac{100}{10}\right)\right] \text{ meq}$$

therefore, buffering capacity of the soil = ($\chi \times 0.01 \times 10$) meq.

or ($0.1 \times \chi$) meq.

***Note:*** The higher the buffering capacity, the larger will be the lime requirement to raise the soil pH to the desired level.

## 3.3 SOIL ACIDITY — CHARACTERISATION FOR ACID SOILS

### 3.3.1 Total Acidity

**Principle**

Total acidity (hydrolytic acidity or titratable acidity) is present in soil in the pH range of 5.5 to 7.0, as hydroxy Al-polymers among acidic soil components. The method is based on the determination of hydrogen ($H^+$) and aluminium ($Al^{3+}$) ions, retained by the adsorptive complex which goes in the solution via exchange, when the soil is treated with a buffered salt solution which undergoes an alkaline hydrolysis. Aluminium also takes part in the formation of hydrolytic acidity.

**Reactions involved**

Extraction: Soil – H + $CH_3$ COO Na → Soil – Na + $CH_3$ COOH

Soil – Al + $3CH_3$ COONa + $3H_2O$ → Soil – Na + $3CH_3$ COOH + $Al(OH)_3$

Precipitate

Titration: $CH_3COOH$ + NaOH    $CH_3COONa$ + $H_2O$

The acidified extract is titrated against a standard alkali to quantitatively estimate the acidity. The acidity arising due to aluminium is not detected in the extract of the alkaline salt solution.

**Reagents**

- 1(N) NaOAC solution adjusted to pH 8.2
- 0.1(N) NaOH solution
- Phenolpthalein indicator

**Procedure**

- Weigh 40 g air-dried soil passed through 2 mm sieve into 250 mL conical flask.
- Add 100 mL of 1(N) NaOAC solution.
- Shake for an hour and filter.
- Titrate the filtrate against 0.1(N) NaOH using 2–3 drops phenolpthalein indicator until a persistent pink colouration is obtained.

**Calculations**

Let

Volume of 0.1(N) NaOH required for sample titration = $V_1$ mL

Therefore, meq of total acidity in 100 mL extract = $0.1 \times V_1$

Thus 40 g soil contains total acidity of $(0.1 \times V_1)$ meq.

Hence, 100 g soil contains total acidity of $[0.1 \times V_1 \times \frac{100}{40}]$ meq

Thus total acidity of the soil = $[0.1 \times V_1 \times \frac{100}{40}]$ meq/100 g

### 3.3.2 Exchange Acidity (Exchangeable Hydrogen and Aluminium)

#### Principle

In acid soils the exchangeable acidic cations are hydrogen and aluminium, the latter of which produces protons by the reaction.

$$Al^{3+} + H_2O \longrightarrow Al(OH)_3 + 3H^+$$

Exchangeable hydrogen and aluminium together constitutes the exchangeable acidity of the soils.

The exchangeable acid cations are extracted from the soil with 1.0 (N) potassium chloride solution. The total exchangeable acidity is determined by titration with standard alkali using phenolpthalein indicator. At the end point pink colour is obtained.

The aluminium is then complexed with potassium fluoride, releasing an equivalent quantity of alkali.

$$Al(OH)_3 + 6KF \longrightarrow K_3AlF_6 + 3KOH$$

The released alkali is titrated with standard acid to measure the exchangeable aluminium.

#### Reagents

- Potassium chloride solution 1(N). Prepare from analytical grade reagent by dissolving 74.56 g in one litre distilled water.
- Potassium fluoride solution 1(N). Dissolve 58 g potassium fluoride in about 900 mL distilled water. Add a few drops of phenolpthalein. If the solution is not pink, add 0.1(N) NaOH in drops till it turns pink. Discharge the colour by adding 0.1(N) $H_2SO_4$ in drops. This eliminates the necessity for blank for the second titration. Dilute to one litre.
- Standard sodium hydroxide solution, 0.1(N)
- Standard sulphuric acid solution 0.1(N)
- Phenolpthalein solution 0.1% in 95% ethanol.

#### Procedure

- Weigh 10 g air-dry sample into a 100 mL conical flask.
- Add 50 mL of the KCl-solution. Mix thoroughly and then allow to stand for 30 min.
- Filter through Whatman No. 40 filter paper into a 100 mL volumetric flask.
- Leach the soil 4 times with 10.0 mL portions of KCl solution.

- Dilute to the volume with the KCl solution.
- Transfer to a 250 mL conical flask, add 6–8 cc phenolpthalein indicator.
- Titrate against the standard NaOH solution to a pink colour that persists for at least 30 seconds.
- Preserve the solution for the second titration.
- Determine the blank correction for the KCl by titrating 100 mL of the KCl solution with the NaOH to a identical end point.
- To determine exchangeable aluminium, add one drop 0.1 N sulphuric acid to the titrated solution, if required to discharge the pink colour, add 10 ml of the KF solution into the flask and mix well.
- Titrate against the standard sulphuric acid till the pink colour disappears.
- Leave aside for ten minutes and continue titrating to a lasting colourless endpoint. This step is followed because the release of the last of the hydroxyl from $Al(OH)_3$ takes place slowly.
- Standardise the NaOH versus Oxalic acid and $H_2SO_4$ versus standard NaOH.

### Calculations

Exchangeable acidity (H + Al), meq/100 g soil

$$= (V_1 - B) \times N_1 \times \frac{100}{W}$$

Exchangeable Al, meq/100 g soil

$$= V_2 \times N_2 \times \frac{100}{W}$$

where $V_1$ = volume of standard NaOH required for sample titration (mL),
B = volume of standard NaOH for blank titration (mL),
$N_1$ = normality of standard NaOH,
$N_2$ = normality of standard acid,
$V_2$ = volume of standard acid required for sample (mL).

## 3.3.3 Extractable Acidity

### Principle

The soil is extracted with a normal ammonium acetate, adjusted to pH = 4.8. Extractable Al in the extract is treated with colour developing aluminon reagent (dilute solution of aurine tri-carboxylic acid) in presence of thioglycollic acid to eliminate the interference of Fe. The intensity of the coloured complex, formed a high temperature is measured colorimetrically at a wavelength of 535 nm. The estimation of extractable acidity or more precisely, extractable Al is an useful index of weathering status of soils and minerals. Extractable aluminium includes exchangeable

Aluminium plus soluble aluminium hydroxide and probably some hydroxy Al-monomers and polymers which perhaps remain strongly adsorbed in the colloidal complex or trapped between the expansible silicate layers of the clay (non-exchangeable Al).

## Reagents

- Normal ammonium acetate adjusted to pH 8.2; Add 58 mL of glacial acetic acid to about 400mL of water in one litre volumetric flask. To it add 70 mL of concentrated $NH_4OH$ through a funnel that extends into the acid and dilute the contents to volume with distilled water. Mix the solution and transfer it to a 2 litre beaker. Adjust the solution pH at 4.8 with 1(N) $CH_3COOH$ (approximately 1 litre $CH_3COOH$ will be required). Mix the contents and keep the solution into a storage flask.
- Aluminon reagent; Dissolve 0.75 g of aluminon and 15 g gum acacia in a beaker with distilled water. Mix thoroughly. Add 342 mL concentrated HCl. Filter in a buchner funnel under suction and dilute the contents to one litre.
- Thioglycollic acid; Dilute 1 mL of pure acid to 100 mL with distilled water.
- Aluminium stock solution; Weigh accurately 0.8094 g $AlCl_3 \cdot 6H_2O$, dissolve with distilled water and dilute to one litre. This is 100 ppm Al solution. Prepare a 10 ppm Al stock from 100 ppm by dilution.

## Procedure

- Weigh 10 g air-dry soil in a beaker and add 50 mL $NH_4OAC$ solution of pH 4.8.
- Mix thoroughly and let stand for 2 hours.
- Filter the suspension through a buchner funnel, using suction.
- Wash the soil in the beaker and on the filter paper with 10 mL of $NH_4OAC$ solution. Repeat 3 to 4 times.
- Dilute the filtrate with $NH_4OAC$ to a volume of 100 mL in a volumetric flask and mix properly by shaking.
- Pipette 5 mL of test solution (containing 1–60 μg of Al)in a 25 mL volumetric flask and add 5 mL of aluminon reagent.
- Dilute the solution to 20 mL with distilled water and add 0.2 mL of thioglycollic acid.
- Mix thoroughly and make up the volume to 25 mL.
- Place the flask in water bath over boiling water for about 10–15 minutes till a brilliant purple or intense red colour is obtained.
- Cool the flask to room temperature and measure the optical density and/or % T in a colorimeter and/or spectrophotometer at an wavelength of 535 nm.
- Prepare a standard curve by pipetting out 0, 1, 2, 4, 8, 12 mL of 10 ppm stock solution into 25 mL volumetric flask and develop colour using the same procedure as in the case of sample preparation.

## Calculations

Let volume of extract pipetted for analysis = V mL

The volume was made up to 25 mL.

$$\text{Therefore, first dilution} = \frac{\text{volume of } NH_4 \text{ OAC added}}{\text{weight of soil}} = \frac{50}{10} = 5 \text{ times}$$

$$\text{Second dilution} = \frac{25}{V} \text{ times}$$

$$\text{Hence, total dilution} = \frac{125}{V} \text{ times}$$

Sample reading from standard curve = S

Concentration of extractable Al from standard curve = C

$$\text{Now extractable Al in soil} = C \times \frac{125}{V} \text{ ppm}$$

$$\text{Extractable Al in soil} = C \times \frac{125}{V} \times \frac{1}{9 \times 10} \text{ meq/100 g}$$

## Non-exchangeable aluminium

This fraction constitutes the hydroxy – Al monomers and polymers which are strongly adsorbed by soil colloid surfaces and behave as if they are virtually non exchangeable.

Non-exchangeable acidity = extractable acidity – exchange acidity.

### 3.3.4 Total Potential Soil Acidity

#### Principle

The total potential soil acidity includes all the acidity components including the weakly acidic R–COOH and R–OH groups on soil organic matter and partially neutralized Al-hydroxy polymers that may be present even in soils of pH >7. The soil is leached with 0.5 (N) $BaCl_2$ and 0.055 (N) triethanolamine buffered at pH 8–8.2. This pH corresponds very nearly to the pH of complete neutralization of soil hydroxy-Al compounds. Barium ion effectively displaces the $H^+$ and $Al^{3+}$ ions and also causes hydrolysis of adsorbed Al-ions and dissociation of acidic R–COOH and R–OH groups, present in soil organic matter, which are neutralized by triethanolamine, which is a weak base. The leachate is titrated with a standard acid using bromocresol green-methyl red mixed indicator. A blank titration is also performed on the same volume of leachate.

#### Reagents

- 1(N) $BaCl_2$ solution
- 0.2 (N) HCl
- Buffer extractant of 0.5(N) $BaCl_2$ plus triethaholamine (pH 8–8.2); Dilute 25 mL of triethanolamine (sp.gr.1.126, normality 8) to 250 mL with water and adjust the pH to 8.0–8.2 with HCl (approx.90 mL 1(N) HCl is required for this partial neutralization process). Dilute the solution to 500 mL and then mix with 500 mL of 1(N) $BaCl_2$ solution. The final solution must be kept free of $CO_2$.

- Bromocresol green–methyl red indicator; Dissolve 0.5 g bromocresol green and 0.1 g methyl red in 100 mL of 95% ethanol and adjust the pH to 4.5.

**Procedure**

- Weigh 10 g soil and take it in a 250 mL conical flask.
- Add 100 mL of extracting buffer solution and shake for half an hour and keep for overnight.
- Transfer the contents to a buchner funnel fitted with a Whatman No. 42 filter paper and carry out gently suction filtration.
- Rinse the conical flask with additional extracting solution so that no soil particle is left over in the flask.
- Now, transfer the leachate to a 250 mL volumetric flask and make up the volume with the extracting solution.
- Pour the leachate into a conical flask and add a few drops of mixed indicator into it. Titrate with 0.2 (N) HCl until the end point colouration (pink) is obtained.
- Perform a blank keeping all conditions identical excepting soil.

**Calculations**

Let

Volume of 0.2(N) HCl required for blank titration = B mL

Volume of 0.2(N) HCl required for sample titration = S mL

Hence, meq. of total potential acidity = (B – S) × 0.2

Now 10 g soil has total potential acidity = [(B – S) × 0.2)] meq

So, 100 g soil has total potential acidity = $\left[\frac{(B-S)\times 0.2\times 100}{10}\right]$ meq

Thus, total potential acidity [2(B – S)] meq/100 g.

### 3.3.5 pH-Dependent Soil Acidity

pH-dependent soil acidity mainly comprises acidity arising from the dissociation of protons from functional groups, viz. carboxyl groups (R–COOH) and phenolic hydroxyl groups (R–OH) on soil organic matter as well as weakly acidic proton on soil mineral edges (OH – groups exposed on broken edges of 1:1 kaolinitic minerals) due to an increase in soil pH.

**Calculation**

pH-dependent acidity = total potential acidity – exchange acidity.

## 3.4 ELECTRICAL CONDUCTIVITY

**Principle**

The conductivity of a soil is precisely the specific conductivity at 25°C of a water extract obtained at a definite soil water ratio. The electrical conductivity is measure on an electrical conductivity bridge and is normally reported in mmhos $cm^{-1}$. A fairly quantitative estimate of the soluble salt content of solutions extracted from the soils can be made from their electrical conductance. Soil extracts obtained using high water to soil ratios are as less accurate measure of the solute content of the soil since more salts may be removed than are ever present in the soil, at field moisture contents. Usually soil: water ratio of 1 : 2.5 or 1.5 is used for routine measurement. Thus the soil: water ratio employed must be specified with the analysis.

The cell constant $k$ of a conductance cell is determined by measurement of the electrical conductance C of a standard KCl solution usually 0.01M KCl solution and use of equation.

$$k = L/C \quad \text{...(3.4.1)}$$

where

L = the known specific electrical conductance mmhos $cm^{-1}$.

C = the conductance of the standard solution measured in a given cell (mmhos).

The measured conductance C of a test solution in mmhos multipled by the cell constant ($k$) gives the specific conductance, L mmhos/cm of the test solution. *i.e.,* L = $k$C

For soil classification purpose the conductivity of saturation extracts of soils is required. However, extraction of solution from a saturated paste is very difficult process. As an approximation, the conductivity of the water extracts from a 1 : 2.5 soil water suspension is determined and the conductivity of the saturation extract is calculated as EC (saturation extract) = EC (1 : 2.5 extract) × 250/saturation percentage.

This does not hold good for soil containing gypsum for which saturation extract must be obtained.

## Reagent and apparatus

- (0.01 M) Potassium chloride prepared from AR grade salt and double distilled water {0.74569 g KCl in 1000 mL or prepare fresh from a stock of 0.2 (M) KCl solution (14.92 g of the salt in 1000 mL)}.
- Conductivity bridge with dip type or pipette type conductivity cells.
- Beakers – 50 mL or 100 mL

## Procedure

- Weigh 20 g air dry soil, into a 100 mL beaker and add 50 mL distilled water.
- Stir at regular intervals for 1 hour.
- Allow to settle for 30 minutes and filter the supernatant through a dry Whatman No. 42 filter paper into a dry beaker.
- Measure the temperature of the soil extract, for future correction.
- Measure the conductivity of the extract with the conductivity bridge. The specific conductivity is obtained by multiplying the electrical conductance with the cell constant. To obtain the conductivity at the temperature of the extract multiply by the appropriate correction factor ($f_t$) obtained from Appendix X.

## Calculations

Electrical conductivity of the extract at 25°C or (Lmmhos/cm) = conductivity of the extract as measured (mmhos) × cell constant ($cm^{-1}$) × temperature correction factor ($f_t$)

$$\% \text{ of salts in soil} = 0.064 \times \text{Lmmhos/cm} \times \frac{\% \text{ of water in soil at extraction}}{100}$$

for (10 g soil + 25 mL water or 20 g soil + 50 mL water) *i.e.,* 1 × 2.5 soil water ratio the above expression will be 0.064 × Lmmhos/cm × 25/10 × 100/100

= 0.064 × Lmmhos/cm × 250/100

= 0.064 × 2.5 × Lmmhos/cm

= 0.16 × Lmmhos/cm

or 0.32 × Lmmhos/cm for 1 : 5 soil solution ratio

## Preparation of saturated soil paste

- Take about 200 g of soil sample (0.2 mm) in a suitable container and add distilled water while stirring with a spatula.
- Tap the container occasionally to consolidate the mass.
- Initially add, sufficient water to bring the soil to near saturation as this gives better results than a slow gradual addition.
- When saturation is attained, the soil surface glistens. The soil flows slightly if the container is tipped and it easily slides off the spatula.
- Allow the paste to stand for one hour and conduct the saturation tests; there should be no free water on the paste surface.
- If extra water has been added mix a little more dry soil.
- With very fine textured soils it is advisable to add the initial water with minimum of stirring to avoid puddling.

## Preparation of saturation extracts

- Spread the saturated paste on a filter paper in a Buchner funnel and filter by suction.
- If gypsum is known to be present, allow the saturated paste to stand overnight.

## Determination of saturation percentage

- Transfer a portion of the saturated soil paste to a tared tin and weigh.
- Dry the paste in an oven at 110°C, cool in a desiccator and weigh again.
- Record the loss in weight

$$\text{S.P.} = \frac{\text{loss in weight} \times 100}{\text{oven dry weight}} \times 100$$

If the air dry moisture content of the soil is known and water added to prepare the paste has been measured, then saturation percentage is calculated from

$$\text{Wt. of oven dry soil} = \frac{\text{Wt. air dry soil}}{(100 + \%\ \text{air dry moisture})}$$

$$\text{Total water} = \text{water added} + (\text{Wt. air dry soil} - \text{Wt. oven dry soil})$$

$$\therefore \quad \text{S.P.} = \frac{\text{Total water} \times 100}{\text{Wt. of oven dry soil}}$$

Alternatively Wilcox (1951) has published formula for calculation of saturation percentage.

$$\text{Saturation moisture percentage} = 37.74\ (2.65V - W)/W - V \qquad \text{...(3.4.2)}$$

where V is the volume in mL of the saturated paste, W is the weight of the saturated paste in gram. The density of water is taken as unity and that of soil particles 2.65. The formula is suitable for minerals soils but not for organic soils.

## 3.5 ORGANIC CARBON

### Principle

The importance of organic carbon estimation lies in the fact that it gives an indication of the organic matter content of the soil which is an important index of soil fertility. The organic carbon content of soil, is reported directly as percentage of C or calculated as organic matter by multiplication with a factor of 1.724 assuming that soil organic matter contains on an average 58 percent carbon, so that 100/58 = 1.724. There is also a close relationship between organic carbon and nitrogen content of the soil (C : N ratio). For organic carbon determination dry combustion and wet digestion methods are used. The former is useful for very accurate purposes since absolute values are obtained, while for routine work Walkley-Black rapid titration method (wet combustion) is extremely useful. Soil organic carbon is oxidized to $CO_2$ in presence of an excess of oxidizing agent such as a mixture of potassium dichromate and sulphuric acid. It is actually the chromic acid formed by the action of potassium dichromate and sulphuric acid which oxidizes the carbon. The excess of dichromate is determined by back titration with ferrous ammonium sulphate (redox titration). The Walkley-Black procedure uses the heat of dilution of sulphuric acid to provide the temperature required. However, this does not enable the dichromate to oxidize all the organically bound carbon. An empirical recovery factor of 0.77 was used by the original authors.

### Reactions

Oxidation of carbon

$$K_2Cr_2O_7 + 4H_2SO_4 \longrightarrow K_2SO_4 + Cr_2\ (SO_4)_3 + 4H_2O + 3O \ ...(\times 2)$$

$$3C + 6O \longrightarrow 3CO_2$$

$$2K_2Cr_2O_7 + 8H_2SO_4 + 3C \longrightarrow 2K_2SO_4 + 2Cr_2\ (SO_4)_3 + 8H_2O + 3CO_2$$

## Titration

$$FeSO_4(NH_4)_2\ SO_4 \cdot 6H_2O \longrightarrow FeSO_4 + (NH_4)_2\ SO_4 + 6H_2O \ ...(\times 2)$$

$$2FeSO_4 + H_2SO_4 + O \longrightarrow Fe_2(SO_4)_3 + H_2O$$

$$2FeSO_4(NH_4)_2\ SO_4 \cdot 12H_2O + H_2SO_4 + O \longrightarrow 2(NH_4)_2\ SO_4 + Fe_2(SO_4)_3 + 3H_2O$$

Action of diphenylamine indicator

$$\underset{\text{diphenylamine}}{2C_6H_5NHC_6H_5} \xrightarrow[-H_2O]{[O]} 2(C_6H_5 \cdot NHC_6H_4) \xrightarrow[-H_2O]{[O]} \underset{\substack{\text{diphenyl benzidine} \\ \text{(violet)}}}{C_6H_5N - C_6H_4C_6H_4N - C_6H_5}$$

## Reagents

- Standard potassium dichromate solution 1(N). Heat, $K_2Cr_2O_7$ in an air oven for 4 hours at 105°C. Dissolve 49.04 of pure $K_2Cr_2O_7$ in distilled water and dilute to one litre.
- Ferrous ammonium sulphate (N/2) approx. Dissolve 196 g ferrous ammonium sulphate, $FeSO_4(NH_4)_2\ SO_4 \cdot 6H_2O$, in water. Add 25 mL of concentrated $H_2SO_4$ and dilute to one litre.
- Red-ox indicator. Use any one of the following;

  *Diphenylamine indicator*: Dissolve 0.5 g of diphenylamine in a mixture of 100 mL conc. sulphuric acid and 20 mL distilled water, and store in a coloured bottle.

  *Ferroin indicator*: Dissolve 1.485 g 1,10 phenanthroline monohydrate in about 80 mL water (warm if necessary and then cool) and added 0.69 g ferrous sulphate heptahydrate. Stir to dissolve and dilute to 100 mL.
- Concentrated sulphuric acid (sp.gr.1.84), 96% concentration or more (for analyzing soil containing chloride, add 15 g silver sulphate per litre).
- Orthophosphoric acid – 85%.

## Procedure

- Grind the soil sample and completely pass through 0.2 mm sieve.
- Take 1.00 g of the soil sample (weighed to the nearest milligram) into a clear, dry 500 mL conical flask.
- Add 10 mL of 1(N) $K_2Cr_2O_7$ by means of a pipette and swirl gently.
- Then add 20 mL of concentrated $H_2SO_4$ rapidly into the solution, and immediately mix by swirling gently at first and then vigorously for a total of one minute.
- Keep the flask on an asbestos pad for 30 minutes.
- Add 200 mL of distilled water, 10 mL of orthophosphoric acid and 1mL of diphenyl amine indicator. A blue violet colour will appear.
- Titrate with (N/2) ferrous ammonium sulphate solution till colour changes from blue violet to green (If ferroin indicator is used the colour change at the endpoint is from blue to red).

- If more than 8 mL of the dichromate solution is consumed repeat the estimation with a smaller quantity (0.25–0.50 g) of the soil sample.
- Simultaneously carry out a blank determination using all the reagents similarly but no soil sample.
- Record the blank value. Since the recovery of organic carbon by this method, varies with the nature of soil and the mean recovery being 77% it is better to express the result as Walkley-Black value.

### Calculations

$$K_2Cr_2O_7 + 4H_2SO_4 = K_2SO_4 + Cr_2(SO_4)_3 + 4H_2O + 3(O)$$

$$C + 2(O) = CO_2$$

1 g mole $K_2Cr_2O_7$ furnishes 3 g atoms of oxygen *i.e.,* 48 g

Considering actually the ion electron reaction occurring;

$$Cr_2O_7^{2-} + 14H^+ + 6e \leftrightarrow 2Cr^{+3} + 7H_2O$$

The equivalent of the oxidant or a reductant is the mole divided by the number of electrons which 1 mole of the substance gains or loses in the reaction.

$$\text{Equivalent} = Cr_2O_7^{2-}/6 = K_2Cr_2O_7/6$$

Also 1 g mole $K_2Cr_2O_7$ represents 6 g equivalent

since 2 g atoms of oxygen *i.e.,* 32 g oxidize 1g atom carbon *i.e.,* 12 g.

∴ 3 g atoms of oxygen will oxidise 12 × 3/2 = 18 g carbon

∴ 6 g equivalent $K_2Cr_2O_7$ can oxidise 18 g carbon

∴ 1 g equivalent $K_2Cr_2O_7$ can oxidise 18 × 1/6 = 3 g carbon

*i.e.,* 1000 mL of (IN) $K_2Cr_2O_7$ = 3 g carbon

or 1 mL of 1(N) $K_2Cr_2O_7$ = 0.003 g carbon

Thus, 1 mL of 1(N) ferrous ammonium sulphate = 1 mL of 1(N) $K_2Cr_2O_7$ = 0.003 g carbon

Alternatively from oxidation number concept

$$\overset{+2\ +12\ -14}{K_2Cr_2O_7} \longrightarrow \overset{+6\ -6}{Cr_2\ (SO_4)_3}$$

The change in oxidation number of two atoms of chromium is from +12 to +6 or by 6 units of reduction. The equivalent of $K_2Cr_2O_7$ is accordingly 1/6 mole. In order to find the equivalent of an oxidizing agent, we divide the mole by the change in oxidation number per molecule. The equivalent of a reducing agent is similarly determined by the change in oxidation number which the oxidised element suffers.

$$\text{Organic Carbon (\%)} \left[\frac{10(B-T)}{B} \times 0.003 \times \frac{100}{w}\right]$$

(Walkley-Black value)

where,

B = volume in mL of ferrous ammonium sulphate solution required for blank titration,

T = volume of ferrous ammonium sulphate (mL) needed for sample titration,

W = weight of soil sample in gram.

More precisely, Organic carbon value considering recovery factor of 0.77 can be calculated using the formula.

$$\left[\frac{10(B-T)}{B} \times 0.003 \times \frac{100}{W} \times \frac{1}{0.77} \times \frac{(100+m)}{100}\right]$$

where $m$ = air dry moisture

% organic matter = % organic carbon × 1.724

### Preparation of primary standard solution

One has to accurately weigh 49.04 g $K_2Cr_2O_7$ (A.R.) and dissolve in one litre distilled water in a volumetric flask. This gives 1(N) $K_2Cr_2O_7$ accurately. But due to human factor sometimes it is not possible to take 49.04 g accurately. The actual amount taken say $x$ gram must be recorded to the nearest milligram. Then strength of $K_2Cr_2O_7$ becomes

$$\frac{\text{Wt. of } K_2Cr_2O_7 \text{ actually taken in 1000 mL}}{49.04} \text{ (N).}$$

### Standardisation of ferrous ammonium sulphate or mohr solution

During blank titration this standardisation process is also performed thus

Let,

- the volume of ferrous ammonium sulphate (N/2 approx.) consumed for blank titration be B mL.
- strength of ferrous ammonium sulphate be S
- volume of $K_2Cr_2O_7$ taken 10 mL
- strength of $K_2Cr_2O_7$ = 1(N) exactly say

$\therefore$ from $V_1S_1 = V_2S_2$ concept of titrimetry;

$$S \times B = 10 \times 1$$

$$\therefore \quad S = \left(\frac{10}{B}\right) \text{ (N)}$$

## 3.6 DETERMINATION OF SOIL MICROBIAL BIOMASS CARBON

An essential and very important component of soil organic matter is the soil microbial biomass which regulates the transformation and storage of nutrients. Since soil micro-organism play a significant role in the retention and release of nutrients, the role of soil microbial biomass must be taken into consideration in any attempt to assess nutrient and energy transfer within soil system.

### *Chloroform Fumigation — Extraction Method**

#### Principle

Microbial constituents released by chloroform fumigation and extracted directly can be used to determine the size of the soil biomass. In the fumigation extraction method, a direct measurement of C and other nutrients contained in the microbial biomass is carried out. Chloroform fumigation is done overnight to kill all the organisms in the soil sample, following which the amount of organic carbon in the sample can be measured by fumigation-extraction method.

#### Reagents

- Distilled chloroform
- Concentrated sulphuric acid
- 0.5(M) $K_2SO_4$: Dissolve 87.1280 g of $K_2SO_4$ in distilled water and dilute to 1000 mL.
- 0.2 (N) $K_2Cr_2O_7$: Dissolve 4.903 g of $K_2Cr_2O_7$ in distilled water and dilute to 500 mL.
- Orthophosphoric acid
- 0.005 (N) Ferrous ammonium sulphate: Dissolve 3.92 g of ferrous ammonium sulphate and 0.15 mL of $H_2SO_4$ in distilled water and dilute to 2 litres.
- Ferroin indicator: Dissolve 1.485 g of orthophenanthroline monohydrate in about 80 mL distilled water (warm if necessary and then cool) and add 0.69 g ferrous sulphate heptahydrate. Stir to dissolve and dilute to 100 mL.

#### Apparatus

- Separatory funnel
- Moisture box
- Vacuum desiccator and vacuum pump
- Rotary shaker
- Hot plate

#### Procedure

- Store soil sample in plastic container which has been sieved through 2 mm mesh to prevent drying. Analyse on the same day.
- For each sample, weigh five sets of 10 g soil.
- Take weight of the empty moisture box.
- Put one set in the moisture box and keep the box in an oven at 100°C for 24 h or until a constant oven dry weight is achieved.

* Readers may consult Jenkinson, O.S. (1988), see suggested reading.

- Cool in a desiccator and weigh the dry soil along with the moisture box. Calculate the moisture content of the soil.
- Out of remaining four sets of soil, keep two sets in 50 mL beakers for fumigation.
- Keep the other two sets after packing in a refrigerator for extraction next day.
- Take 20 mL of chloroform in a separatory funnel for each 10 g of soil sample. Wash the chloroform two times with concentrated $H_2SO_4$ (each with half the volume of chloroform); shake well and discard the acid (lower phase) after phase separation. Open the stop cock very carefully after each shaking to release pressure.
- Wash twice with equal volume of distilled water similarly to make the chloroform free of ethanol and collect the bottom whitish phase.
- Keep to the ethanol-free chloroform (not exceeding 40 mL) in 100 mL beakers containing some glass beads to avoid bumping.
- Place all beakers containing soil and chloroform in a vacuum desiccator, the inner surface of which is lined with moistened filter paper. Do not use plastic desiccator. Properly seal the lid-joint using high density vacuum grease. Use a rubber tube to direct the exhaust through water.
- Switch on the vacuum pump and keep it on until the chloroform boils for about 5 minutes. Close the outlet and place the desiccator in dark for 24 hours.
- After completion of 24 hours release the vacuum, take out the beaker containing chloroform and inner paper lining. Perform back suction five to six times in order to remove any excess/adhered chloroform vapour. Release vacuum slowly.
- Take out the unfumigated soil sample from refrigerator and thaw it.
- Transfer both the fumigated and unfumigated soil samples in 250 mL conical flask. Add 25 mL of 0.5 (M) $K_2SO_4$ and shake for ½ hour on a reciprocal shaker.
- Pipette out accurately 10 mL of the filtrate in 500 mL conical flask. Add 2 mL of 0.2 (N) $K_2Cr_2O_7$; 10 mL of concentrated $H_2SO_4$ and 5 mL of orthophosphoric acid to each flask.
- Perform two blanks with 10 mL of distilled water and with the reagents mentioned above.
- Keep the flasks on hot plate at 100°C for ½ hour under refluxing condition. Remove the flasks and add about 250 mL of distilled water immediately and allow the contents to cool down to room temperature.
- Add 2 to 3 drops of ferroin indicator and titrate the contents against 0.005 (N) ferrous ammonium sulphate to obtain a brick-red end point.

## Calculations

**I. Soil water content ($S_w$)**

$$S_w\ \% = \frac{\text{Wt. of wet soil (g)} - \text{Wt. of oven dry soil (g)}}{\text{Wt. of oven dry soil (g)}} \times 100$$

**II. Weight of soil (oven dry weight equivalent) taken for microbial biomass measurement ($W_s$).**

$$W_s\ (g) = \left(\frac{\text{Wt. of wet soil (g)}}{\{100 + S_w(\%)\}}\right) \times 100$$

**III. Total volume of solution in the extracted soil ($V_s$)**

$$V_s\ (mL) = \text{Wt. of wet soil (g)} - \text{Wt. of oven dry soil (g)} + \text{extractant volume (mL)}.$$

**IV. Determination of extractable carbon (EC in mg $mL^{-1}$)**

- Standardisation of Ferrous ammonium sulphate (FAS) solution.

$$\text{Strength of FAS (N)} = \frac{\text{volume of } K_2Cr_2O_7 \text{ (2 mL)} \times \text{strength of } K_2Cr_2O_7}{\text{mean titre value for blank(mL)}}$$

- Volume of $K_2Cr_2O_7$ consumed by Ferrous ammonium sulphate in any sample

$$Y\ (mL) = \frac{\text{normality of FAS} \times \text{titre value (mL)}}{\text{normality of } K_2Cr_2O_7}$$

- Volume of $K_2Cr_2O_7$ consumed for oxidizing easily mineralisable C in 10 mL of extractant = (2 – Y) mL
- Extractable C (EC) in μg $mL^{-1}$

$$1\text{ mL } 1\text{ (N) } K_2Cr_2O_7 \equiv 0.003\text{ g of C}$$

$$1\text{ mL } 0.2\text{ (N) } K_2Cr_2O_7 \equiv (0.003 \times .2)\text{ g of C}$$

$$(2-Y)\text{ mL } 0.2\text{ (N) } K_2Cr_2O_7 \equiv \{0.003 \times .2 \times (2-Y)\}\text{ g of C}$$

$$= \{600 \times (2-Y)\}\ \mu g\text{ of C}$$

Therefore, the amount of extractable carbon (EC)

$$EC\ (\mu g\ mL^{-1}) = \left\{\frac{600\,(2-Y)}{10}\right\}$$

- Total weight of extractable C in the fumigated ($EC_f$) and unfumigated ($EC_{uf}$) soil samples.

$$EC_f \text{ or } EC_{uf}\ (\mu g\ g^{-1}\text{ soil}) = \frac{EC(\mu g\ mL^{-1}) \times V_s(mL)}{W_s(g)}$$

- Microbial biomass carbon in soil (MB – C)

$$MB - C\ (\mu g\ g^{-1}\text{ soil}) = \frac{(EC_f - EC_{uf})}{K_{ec}}$$

where $K_{ec} = 0.35$ (Voroney *et al.*, 1991) represents the efficiency of extraction of organic carbon and its value depends on physical and chemical properties of soil.

*Voroney, R.P., and E.A. Paul, 1984. Determination of $K_c$ and $K_n$ *in situ* for calibration of the chloroform fumigation incubation method. *Soil Biol. Biochem.* **16**: 9–14.

## 3.7 TOTAL NITROGEN (MODIFIED KJELDAHL'S METHOD)

### Principle

Most of the nitrogen in soils occur in organic form. The available forms – ammonium and nitrate occurs ordinarily in relatively small amounts. The Kjeldahl method includes both organic and ammonium forms and with modifications includes the nitrate form. The nitrate nitrogen must be included in total, nitrogen determination of soils containing appreciable amounts. Organic and nitrate nitrogen is converted to ammonium sulphate during digestion with concentrated $H_2SO_4$. A digestion mixture is added to speed up the reaction. Sodium or potassium sulphate raises the boiling temperature of the acid to provide more effective oxidation while the reaction is catalysed by addition of copper sulphate-selenium mixture. The salt catalyst mixture prescribed is one of the most effective, mercuric oxide or mercuric salts are sometimes used in place of selenium but are inconvenient as the mercuric ion must be precipitated before distillation to prevent ammonia being trapped as the mercury ammonium complex, this precipitation is usually effected as the sulphide and leads to objectionable odour in the distillate and sometimes black deposits in the condenser. In modified method concentrated sulphuric acid-salicylic acid is used along with sodium thiosulphate and mossy zinc/zinc dust. During digestion nitrate loss generally occurs as $HNO_3$ (if sample contains appreciable nitrate). Hence, salicylic acid is used which fixes the nitrate to nitro-salicylic acid. Addition of zinc dust or sodium thiosulpate reduces the nitro-salicylic acid to amino-salicylic acid which on digestion with concentrated $H_2SO_4$ is converted to ammonium sulphate.

### Reactions involved

$$\text{Organic nitrogenous compounds} \xrightarrow[\text{Catalyst}]{H_2SO_4} (NH_4)_2SO_4$$

$$\underset{\text{(Salicylic acid)}}{C_6H_4OH \cdot COOH} + HNO_3 \longrightarrow \underset{\text{(Nitro-salicylic acid)}}{C_6H_3OH \cdot NO_2 \cdot COOH} + H_2O$$

The nitro-salicylic acid is reduced to amino-salicylic acid by the action of sulphurous acid formed by interaction of sulphuric acid and sodium thiosulphate.

$$Na_2S_2O_3 + H_2SO_4 \longrightarrow Na_2SO_4 + \underset{\text{(Sulphurous acid)}}{H_2SO_3} + S$$

$$\underset{\text{(Nitro-salicylic acid)}}{C_6H_3OH \cdot NO_2 \cdot COOH} + 3H_2SO_3 + H_2O \longrightarrow \underset{\text{(Amino-salicylic acid)}}{C_6H_3OH \cdot NH_2 \cdot COOH} + 3H_2SO_4$$

$$\underset{\text{(Amino-salicylic acid)}}{2C_6H_3OH \cdot NH_2 \cdot COOH} + 27H_2SO_4 \longrightarrow \underset{\text{(Ammonium sulphate)}}{(NH_4)_2SO_4} + 26SO_2\uparrow + 14CO_2\uparrow + 3H_2O$$

The ammonium sulphate on distillation with NaOH gives ammonia

$$(NH_4)_2SO_4 + 2NaOH \longrightarrow Na_2SO_4 + 2NH_3 \uparrow + 2H_2O$$

The liberated ammonia is absorbed in a known volume of standard sulphuric acid solution containing methyl red indicator

$$2NH_3 + H_2SO_4 \longrightarrow (NH_4)_2SO_4$$

The acid used for absorbing ammonia is determined by back titration with standard solution of sodium hydroxide.

***Note:***

- The determination of total nitrogen is not as simple as it is often thought to be. It is subject to many difficulties, any of which may lead to low results. For example, an hour of digestion may be necessary after the digest turns clear, to release all of the nitrogen, the most important consideration being catalyst selected and the digestion temperature. If the digestion temperature is too low (below 360 °C), the release is slower and/or incomplete and if too high (over 410 °C), some loss of $NH_3$ from the mixture results.
- The granulated zinc is added in order to prevent bumping during distillation. Zinc reacts with sulphuric acid to produce minute bubbles of hydrogen which aid in preventing bumping. The sulphuric acid used may contain traces of ammonium sulphate and the distilled water exposed to the laboratory air may contain traces of ammonium hydroxide. Hence, it becomes necessary to carry out a blank determination by which any extraneous nitrogen is determined and subtracted from the total value.
- The C : N ratio is obtained by dividing the percentage of organic carbon by the percentage of total nitrogen.

## Reagents

- Sulphuric acid – Salicylic acid mixture ; Dissolve 1 g salicylic acid in 30 mL of concentrated sulphuric acid
- Sodium thiosulphate ($Na_2S_2O_3 . 5H_2O$)
- Digestion mixture (Mix 10 parts of potassium sulphate, 1 part of copper sulphate and 0.5 part of selenium powder)
- Standard Sodium hydroxide 0.1(N)
- Concentrated sulphuric acid
- Standard sulphuric acid 0.1(N)
- Zinc dust
- Methyl red indicator
- Sodium hydroxide (450 g/L)

## Procedure

## Digestion

- Take 10 g of soil sample (passed through 0.5 mm sieve) in a Kjeldahl flask (600–800 mL capacity).

- Add 35 mL of the sulphuric-acid salicylic-acid mixture.
- Shake and let stand for 30 minutes for nitrates to react with the salicylic acid. Then add 5 g of $Na_2S_2O_3.5H_2O$ (or alternatively 2 g Zn dust) and heat gently on low flame for 5 minutes taking care to avoid frothing.
- Cool and add 10 g of digestion mixture and continue digestion gradually raising the temperature until the solution becomes clear and acquires a grayish blue or greenish colour (usually a total of about 2–3 hours is required for digestion).
- Cool and add slowly, with intermittent shaking 300 mL of distilled water. This solution is further cooled (heat of dilution), (If large quantities of sand are present bumping during distillation is sometimes severe. This can be avoided by washing the acid solution into another flask, the sand being left in the original flask).
- Now fit to the distillation apparatus.
- Add 100 mL of concentrated sodium hydroxide (usually 40%) and several pieces of granulated zinc.
- Add 1 teaspoon glass beads.
- Connect, to the distillation head and distill off 150 mL into 25 mL of standard $H_2SO_4$ solution 0.1(N) containing methyl red indicator.
- Back titrate the excess acid with standard alkali *i.e.,* 0.1(N) NaOH.
- Perform a blank in the exactly same manner without the soil.

### Standardisation Technique

Standardisation of $H_2SO_4$ (0.1 N) approx is generally performed by titrating against standard $Na_2CO_3$ solution as primary standard (equivalent weight of $Na_2CO_3$ is 53; hence, for 0.1(N), accurately weigh 5.3 g $Na_2CO_3$ in a 250 mL volumetric flask and make up the volume. This gives 0.1N exact $Na_2CO_3$ solution). If exact weight is not taken then record the weight to the nearest milligram which is actually taken. Pipette out 25 mL of $Na_2CO_3$ solution and add 1–2 drops methyl orange (or methyl red) indicator when solution turns yellow. Titrate with 0.1(N) $H_2SO_4$ taken in the burette until the colour becomes faintly yellow, continue addition of acid carefully, drop wise stirring until the colour of solution just assumes faint pink or orange shade. Note the end point reading from the burette in mL. Repeat thrice.

$$\text{Strength of } H_2SO_4 = \frac{\text{volume of } Na_2CO_3 \times \text{strength of } Na_2CO_3}{\text{volume of acid}}$$

Hence, exact strength of $H_2SO_4$ is known. Likewise 0.1N approx. NaOH is standardised against standard $H_2SO_4$. Pipette out 25 mL of 0.1 N approx NaOH solution into a beaker. Add 1 drop phenolphthalein indicator, the solution turns pink. Titrate with standard $H_2SO_4$ solution from burette with continual stirring. The pink colour becomes fainter and fainter until it is very faint. Continue adding acid dropwise until with one drop the pink colour just disappears. Note the volume of acid added from burette in mL at the end point. Repeat the experiment thrice.

$$\text{Strength of NaOH solution} = \frac{\text{volume of standard acid} \times \text{exact strength of acid}}{\text{volume of NaOH solution}}$$

Then exact strength of NaOH solution is thus known.

The actual strength of $Na_2CO_3$ then is calculated as

$$\text{Strength of } Na_2CO_3 = \frac{\text{weight of } Na_2CO_3 \text{ actually taken in one litre}}{5.3} = \left(\frac{N}{10}\right)$$

However, 0.1(N) approx NaOH can also be standardised versus 0.1(N) accurately prepared oxalic acid which is a primary standard using phenolpthalein indicator. Equivalent weight of Oxalic acid ($H_2C_2O_4 \cdot 2H_2O$) is 63 [63 g in one litre gives 1(N) or 6.3 g in 1 litre gives 0.1(N)]. Therefore 1.575 g is to be weighed accurately in 250 mL volumetric flask and volume made up to the mark with distilled water to prepare 250mL of 0.1(N) oxalic acid solution. Then a known volume say 25 mL of this acid will be taken by pipette plus (1 drop phenolpthalein) and volume of NaOH required to neutralise from burette is noted in mL at the end point when pink colour just appears with one drop of NaOH.

Again;

volume of NaOH × strength of NaOH = volume of oxalic acid × strength of oxalic acid.

Hence exact strength of NaOH can be evaluated since all other values are known.

### Calculations

Refer Article No.3.8

## 3.8 MINERALISABLE NITROGEN

### Principle

The term available nitrogen of soil includes all ammonical and nitrate nitrogen, the forms in which plants are known to absorb nitrogen from soil. More than 90% of soil nitrogen normally exists in complex combination with the organic matter *i.e.,* the humus fraction. It becomes available to crops after breakdown to simple from followed by mineralization. This type of transformation is mostly biological in nature. Mineralization of soil nitrogen generally refers to the conversion of organic nitrogen to inorganic nitrogen and in practice refers to production of ammonium and nitrate and other inorganic forms that are usually transitional, although for most accurate results nitrite should also be considered. Mineralization is largely microbiological process, the organic nitrogen being first changed to ammonium and then via nitrite to nitrate nitrogen, (nitrification process).

Attempts to measure 'available nitrogen' by purely chemical means have mostly been approximate because it is almost impossible to imitate the biological decomposition process. Hence, methods involving the determination of mineral forms of nitrogen ($NH_4$–N and $NO_3$–N) produced on incubation of soil under aerobic and anaerobic conditions are generally followed. The chemical method for determination of available soil nitrogen involve measurement of a fraction of easily hydrolysable nitrogen by using solutions of dilute acids or alkali. Distillation with alkaline potassium permanganate solution has often been adopted for estimating the readily oxidisable and reactive forms of soil nitrogen. In the alkaline permanganate method (Subbiah and Asija, 1956) nitrogen is released by the alkaline permanganate solution and estimated by the usual ammonia distillation procedure, the distillate being absorbed in standard acid and excess acid back titrated with standard alkali using methyl red indicator.

### Reagents

- 0.32% potassium permanganate solution — freshly prepared
- 2.5% sodium hydroxide solution — freshly prepared
- Standard sulphuric acid 0.02(N)
- Standard sodium hydroxide 0.02(N).
- Methyl red indicator (Dissolve 1 g methyl red in 200 mL of rectified spirit)
- Liquid paraffin (extra pure)
- Glass beads

### Procedure

- Weigh accurately 20 g of the soil sample in a distillation flask.
- Add 20 mL of distilled water, 100 mL of potassium permanganate solution and 100 mL of 2.5 per cent sodium hydroxide solution (the frothing during boiling is prevented by adding liquid paraffin (1mL) and bumping by adding a few glass beads.
- Immediately after alkali addition connect to the distillation apparatus and distill the contents in Kjeldahl assembly at a steady rate.
- Pipette out 25 mL of standard sulphuric acid (0.02 N) in a conical flask.
- Add methyl red indicator and dip the end of delivery tube in it.
- Distil the ammonia gas from the distillation flask for about 30–40 minutes until distillation is completed and collect about 100 mL of the distillate.
- Back titrate the excess acid with standard alkali *i.e.,* 0.02 N NaOH. (Colour change at end point is usually pink to faint yellow or straw).
- Perform a blank without sample.
- Standardise 0.02 N approx. NaOH against accurately prepared oxalic acid 0.02 (N) (by taking exact weight or to the nearest milligram) using phenolpthalein indicator. Hence, exact strength of NaOH is known from $V_1S_1 = V_2S_2$ relationship of acidimetry and alkalimetry.

### Calculations

$$\% \text{ N in soil} = (S-T) \times N \times \frac{1.4}{w}$$

where, S = blank titration, mL standard NaOH required for 25 mL $H_2SO_4$ used for receiving the distillation of blank.

T = titration of sample, mL standard NaOH required for 25 mL $H_2SO_4$ receiving the sample

N = Normality of standard alkali,

$w$ = sample weight in gram

### ***Fundamental basis for calculation of Nitrogen***

### Reactions involved

$$2NH_3 + H_2SO_4 \longleftrightarrow (NH_4)\ SO_4 \qquad ...(3.8.1)$$

From (3.8.1) it is evident, 1 g mole $H_2SO_4$ reacts with 2 g mole $NH_3$

$$\therefore \qquad H_2SO_4 \equiv 2NH_3 \equiv 2\ N \qquad ...(3.8.2)$$

But 1 g mole $H_2SO_4$ = 2 g equivalent $H_2SO_4$ ...(3.8.3)

(Because equivalent weight of $H_2SO_4$ = molecular wt./2)

From equations 3.8.2 and 3.8.3 it follows;

$$1 \text{ g equivalent } H_2SO_4 \equiv \frac{1}{2} \text{ g mole } H_2SO_4 \equiv 1 \text{ g mole } NH_3 \equiv 1 \text{ g mole N} \qquad ...(3.8.4)$$

Also 1000 mL 1(N) $H_2SO_4$ contains 1 g equivalent $H_2SO_4$

Hence, 1000 mL 1(N) $H_2SO_4 \equiv$ 17 g $NH_3$ = 14 g N ...(3.8.5)

or 1 mL 1(N) $H_2SO_4 \equiv$ 0.017 g $NH_3 \equiv$ 0.014 g N ...(3.8.6)

### Alternative calculation with respect to standard $H_2SO_4$

Let exact strength of oxalic acid, obtained from weight = $x$ (N)

Hence, from oxalic acid vs NaOH, titration

we get exact strength of NaOH, say $y$ (N)

Again from standard NaOH vs $H_2SO_4$ titration, we get exact strength of $H_2SO_4$, say $z$(N)

Also, say 25 mL of standard $H_2SO_4$ was taken to receive distillation of blank

and

25 mL of standard $H_2SO_4$ was taken to receive distillation of sample

Suppose $a$ mL of standard NaOH was required for sample titration and $b$ mL of standard NaOH was required for blank titration.

From standard NaOH vs standard $H_2SO_4$ titration we get,

$p$ mL standard NaOH $\equiv$ $q$ mL standard $H_2SO_4$

$\therefore$ $a$ mL standard NaOH = $(q/p \times a)$ mL standard $H_2SO_4$ = T mL say (from sample titration)

Also,

B mL standard NaOH = $(q/p \times b)$ mL standard $H_2SO_4$ = B mL say (from blank titration)

Hence, actually the mL standard $H_2SO_4$ consumed by $NH_3$ = (B−T) mL

Now, 1 mL 1(N) $H_2SO_4$ = 0.014 g N

Therefore, (B – T) mL $z$(N) $H_2SO_4$ ≡ [0.014 × (B–T) × $z$] g N

$w$ g soil sample contains [0.014 × (B–T) × $z$] g N

$$\therefore\ 100 \text{ g soil sample will contain } \frac{[(B-T)\times z\times 0.014]}{w}\ \%\ \text{Nitrogen}$$

$$\text{Hence N percentage} = \frac{[(B-T)\times z\times 0.014]}{w}$$

## 3.9 DETERMINATION OF SOIL MICROBIAL BIOMASS NITROGEN

### Procedure

- Follow the steps (1–15) in Article No. 3.6 in the determination of soil microbial biomass carbon.
- Determine total N in the extract by Kjeldahl digestion as outlined in Article No. 3.7.

### Calculations

Microbial biomass nitrogen (MB – N)

$$\text{MB} - \text{N (mg g}^{-1}\text{ soil)} = \frac{EN_f - EN_{uf}}{K_{en}}$$

where $N_f$ = total N from the fumigated soil extract,

$N_{uf}$ = total N from the unfumigated soil extract,

$K_{en}$ = efficiency of extraction of organic microbial N and inorganic N from soil and usually varies from 0.54 to 0.62 (Jenkinson; 1988).

## 3.10 TOTAL PHOSPHORUS

The total content of the elemental phosphorus in soils can be extracted and determined by perchloric acid digestion followed by spectrophotometric determination.

### $HClO_4$ digestion for total P

- Weigh 2 g sample of soil accurately (5.0 g for soil low in phosphorus) which has been passed through 0.5 mm sieve in a 250 mL conical flask (If the sample is high in organic matter, add 20 mL $HNO_3$ and heat on a steam plate to effect preliminary oxidation. For soil, fairly low in organic matter, the $HNO_3$ treatment is omitted).
- Add 30 mL of 60 percent $HClO_4$ and carry out digestion at 130°C in a digestion chamber which effectively removes $HClO_4$ fumes.
- Carry out the digestion until the solution appears colourless with slight increase in temperature if necessary. (Generally, 50 minutes is sufficient for digestion). As the digestion becomes complete dense white fumes of $HClO_4$ appear and the silica becomes white.

---

Jenkinson, D.S., 1988. The determination of microbial biomass carbon and nitrogen in soil pp. 368–386. In J.R. Wilson (ed.). Advances in nitrogen cycling in agricultural ecosystems. C.A.B. International, Wallingford.

- When the digestion is completed, remove the flask. Cool sufficiently and add 50 mL of double distilled water.
- Transfer the solution through a filter to a 250 mL volumetric flask.
- Wash the residue to bring the volume of solution up to the mark.
- This is the stock solution of total P.
- Take an aliquot from this stock or after dilution of the initial stock if necessary and determine the 'P' spectrophotometrically as usual at 660 mμ.

## 3.11 EXTRACTABLE PHOSPHOROUS DETERMINATION – GENERAL DISCUSSION

The phosphate ion concentration in soil solution is essentially controlled by the heterogeneous equilibria of the form;

Solid phase P $\rightleftharpoons$ Soil solution P $\rightleftharpoons$ Precipitated P.

Inorganic orthophosphate ions viz. $H_2PO_4^-$, $HPO_4^{2-}$ and $PO_4^{3-}$ existing in soil solution are the plant available form, the most accessible ion being $H_2PO_4^-$ with largest activity coefficient followed by $HPO_4^{2-}$. Of the factors which control the availability of the inorganic soil phosphorus, soil pH is of primary importance. It is well established fact that at relatively low pH, $H_2PO_4^-$, ions usually exists. On decreasing acidity, first the $HPO_4^{2-}$ and then $PO_4^{3-}$ ions play the dominant role. Thus at intermediate pH values any two of the three aforesaid ions coexists. In general the availability of these ions to plants is considered to follow the order $H_2PO_4^- > HPO_4^{2-} > PO_4^{3-}$ but the presence of iron and aluminium at low pH and calcium at high pH vitiates this sequence. For soils having pH in the range from 4.5 to 7.5, ions of $H_2PO_4^-$, as well as $HPO_4^{2-}$ exist in soil solution. At pH > 8.3 $HPO_4^{2-}$ ions predominate in solution. However at pH > 9, the $PO_4^{3-}$ ion becomes more important than $H_2PO_4^-$.

Several chemical extractants has been developed over the years from the concept of simulating plant root activity and with the purpose to obtain an appropriate measure of the readily soluble inorganic forms to represent the plant available soil P. Out of the several extractants, depending upon soil pH, presence of iron and aluminum compounds or calcium carbonate; one particular method is selected for analysis. The most common methods along with extractant composition, soil extractant ratio and shaking time are furnished below:

| Methods | Extractant composition | Soil extractant ratio | Shaking time (min) |
|---|---|---|---|
| Morgan's method | 10 mL of glacial acetic acid in 1 litre of 0.5% NaOH adjusted to pH 8.4. | 1 : 5 | 30 |
| Mehlich's method | 0.05 (N) HCl + 0.025 (N) $H_2SO_4$ (pH 1.2) – prepared by 4.3 mL conc. HCl + 0.7 mL conc. $H_2SO_4$ in one litre. | 1 : 4 | 5 |
| Bray's I method | 0.03 (N) $NH_4F$ + 0.025 (N) HCl (pH 3.5) prepared by dissolving 1.11 g $NH_4F$ in 2.1 mL of conc. HCl in one litre | 1 : 10 | 5 |

| Methods | Extractant composition | Soil extractant ratio | Shaking time (min) |
|---|---|---|---|
| Bray's II method | 0.03 (N) $NH_4F$ + 0.1 (N) HCl (pH 1.0) prepared by dissolving 1.11 g $NH_4F$ in 8.5 mL conc. HCl in one litre. | 1 : 20 | 2/3 |
| Truog's method | 0.002 (N) $H_2SO_4$ (pH 3) prepared by diluting 0.2 (N) acid one hundred times adding 3 g $K_2SO_4$ per litre of extractant. | 1 : 100 | 30 |
| Olsen's method | 0.5 (M) $NaHCO_3$ (pH 8.5) prepared by dissolving 42 g $NaHCO_3$ per litre and adjust the pH to 8.5 with NaOH. | 1 : 20 | 30 |

However, based on higher correlation of soil test with crop response, dilute acid fluoride extraction (Bray's No.1 extractant) or $NaHCO_3$ adjusted at pH 8.5 (Olsen's extractant) considering all soils including acid, neutral, alkaline and calcareous are most commonly used for routine soil testing and soil fertility evaluation programme.

After extraction, in the filtered extract, phosphorus is estimated colorimetrically by adding ammonium molybdate and thereafter reducing the molybdenum – phosphate complex in acidic medium with a reducing agent for which stannous chloride is used. The intensity of the blue colour molybdenum blue is directly related to the quantity of orthophosphate ion and thus provides a measure for the concentration of P in test solution. The absorbance or transmittance is measured spectrophotometrically at 660 mμ. wavelength.

The molybdenum blue colour develops at a rate depending on the temperature and the stannous ion concentration. The colour is stable for a limited period of time and then begins to fade somewhat rapidly. Hence the readings must be taken within the period of stability generally between 5 and 6 minutes after $SnCl_2$ addition but definitely before 15 minutes.

The fundamental principle governing the interaction of ammonium molybdate with phosphorus is complexation mechanism where hetorpolycomplexes are thought to be formed by coordination of molybdate ions with phosphorus as the central coordinating atom, the oxygen of the molybdate radicals being substituted for that of phosphate.

$$H_3PO_4 + 12H_2MoO_4H_3P(Mo_3O_{10})_4 + 12H_2O \quad ...(3.11.1)$$

Ions besides ($P^{5+}$) which may act as the central coordinating atom to form 12-fold heteropoly acids with molybdate include aresenic ($As^{5+}$), silicon ($Si^{4+}$), germanium ($Ge^{4+}$) and under some conditions molybdenum ($Mo^{6+}$) and boron ($B^{3+}$). Tungstate can also be coordinated about P as central atoms but with less avidity. The heteropolycomplexes, before reduction give a yellow hue to their water solution. With high P concentrations, a yellow precipitate is formed. In solution of low enough concentration to be suitable for determination by reduction to form the blue colour, the yellow colour is so faint that, it is not noticed and spectrophotometric measurements is done without any problem. The molybdenum blue colour is produced when either molybdate or its hetropolycomplexes are partially reduced. Some of the molybdenum ions are reduced from +6 to a low valence state, probably +3 and/or +5, involving unpaired electrons due to which spectrophotometric resonance (blue colouration) would be expected.

***Note:*** Arsenate often can cause high results, because an arsenate molybdate acid mixture is also reduced to molybdenum blue and hence soils contaminated with arsenate from arsenious plant sprays or due to other geologic reasons cannot be treated as per the method specified.

The molybdo-arsenic acid blue colour is usually excluded from the phosphorus analysis by reduction to arsenious acid prior to the addition of ammonium molybdate to form the heteropoly complex. The complex does not form with arsenious radical. Moreover the P can be precipitated with $Al(OH)_3$ and the precipitate treated with HF, HBr, HCl and $H_2SO_4$ to volatilize As, Ge and Si which may be co-precipitated leaving P for analysis.

**Optimum reagent concentration:** The optimum concentration of acid molybdate and reductant is that which will give the maximum of colour per unit of P present in accordance with Beer's law, and the minimum of fading.

### 3.11.1 Ammonium Fluoride-Hydrochloric Acid Extractable Phosphorus of Soils: Bray's 1 Method; Bray Kurtz, 1945 (For Soils with pH Around to 5.5 or Less)

#### Principle

Fluoride ion ($F^-$) from ammonium fluoride complex with aluminium ($Al^{+++}$) and ferric ($Fe^{+++}$) in acid solution in the form of double fluoride with consequent release of phosphorus held in the soil by these trivalent ions

$$2NH_4F + 3HF + AlPO_4 \longrightarrow H_3PO_4 + (NH_4)_3AlF_6$$

$$3NH_4F + 3HF + FePO_4 \longrightarrow H_3PO_4 + (NH_4)_3FeF_6$$

Thus in contact with aluminium and ferric phosphates, the aluminium and ferric ion concentrations are reduced and so the phosphate ion concentration is increased to maintain the solubility product at their constant levels. $AlPO_4$ represents various hydrated and hydroxyl phosphates of aluminium, including any adsorbed or precipitated surface layers on oxides and alumino-silicates. $FePO_4$ similarly, represents various hydrated and hydroxyl phosphates of iron including adsorbed or precipitated surface layers on iron oxide.

Very short shaking period is advocated (one minute or 40 seconds for Bray 2 extractant). In fact, for may soils, there is little change in the phosphate concentration of the extract with increase of time but for some soils either more or less phosphate is extracted. Moreover, the fluoride ion has a slightly depressant effect on molybdenum blue colour development and hence its concentration must be kept constant in the test and standard.

#### Reagents

- Double distilled water.
- Bray's No. 1 reagent (0.03 N $NH_4F$ and 0.025 N HCl); 1.11g solid $NH_4F$ (AR/LR) and 4.16 mL of 6(N) HCl per litre. This extractant may be stored in a glass container for a year without appreciable deterioration.

- Dickman and Bray's chloromolybdic acid reagent, 1.5% (For Bray's No.1 method), Weigh exactly 15.0 g of ammonium molybdate (A.R)[$(NH_4)_6 Mo_7O_{24} \cdot 4H_2O$]. Dissolve in about 300 mL double distilled water and warm to about 50 °C. Filter the solution to remove sediments if necessary. Cool the molybdate solution and to this add 350 mL 10(N) HCl slowly with rapid, stirring. Allow to cool at room temperature and dilute with double distilled water to 1000 mL in a one litre volumetric flask, mix thoroughly. Store in an amber, glass stoppered bottle. (Alternatively carbon paper may be wrapped on all sides of the volumetric flask and tied with rubber bands) This is a 1.5% ammonium molybdate solution in 3.5 N HCl. (in every occasion the normality of the HCl must be checked against standard NaOH by titration).
- Stannous chloride solution; Dissolve 10 g of crystalline stannous chloride by warming and store in an amber coloured bottle, carefully avoiding contact with air. This is 40% stock solution of $SnCl_2$. Just before use prepare freshly diluted stannous chloride. Pipette 1 mL 40% $SnCl_2$ into 330 mL of double distilled water (A piece of tin metal (AR) added to the stock solution will preserve the stock solution for a long time).

### Procedure

- Weigh exactly 5.00 g soil in a 100 mL conical flask and add 50 mL of Bray's No.1 reagent with the help of pipette.
- Stopper the flask with rubber cork and shake for 5 minutes.
- Filter the suspension immediately through Whatman No. 42 filter paper (Discard 1–2mL of initial filtrate by the way of rinsing the container in which the filtrate is collected).

*Note:* To avoid interference of fluoride, 7.5 mL of 0.8 (M) Boric acid (50 g $H_3BO_3$ per litre) may be added to 50 mL of extract if necessary [$4F^- + H_3BO_3 + 3H^+ \rightarrow (BF_4)^- + 3H_2O$]. Neither fluoroborate nor basic acid interfere Fluroborate is very slightly ionized.

### Colour development technique

- Pipette 5 mL of soil extract into a 25 mL volumetric flask and to it add 5 mL of chloromolybdic acid reagent (1.5%). Wash the neck of the flask with double distilled water until the contents are diluted to about 22 mL.
- Finally, add 1mL dilute stannous chloride and make up the volume up to the mark with double distilled water.
- Measure the intensity of the blue colour, just after 4–5 minutes, spectrophotometrically at 660 mμ (red filter) and determine the concentration of P from the standard curve.
- With each set of samples perform a blank (without soil).

Standard curve construction for P is discussed later (Article No. 3.11.2)

## 3.11.2 Alkaline Extraction of Soil Phosphorus (Olsen's Method)

### Principle

Alkaline and neutral soils mainly contain di- and tri-calcium phosphates as insoluble phosphates whereas acid soils are preponderant in aluminium and ferric phosphates as insoluble forms.

Phosphate ion ($HPO_4^-$, $H_2PO_4^-$) are however present in soil solution according to the relative amounts of calcium, aluminium and ferric ions. If the concentration of the metallic ions are reduced the concentration of the phosphate ions increases in order to maintain the various solubility products at their constant values. An alkaline (pH 8.5) bicarbonate solution can repress the concentration of calcium ions by precipitation as calcium carbonate and of aluminium and ferric ions as hydroxides. Thus phosphate ions concentrations are increased and available phosphate can be extracted from soil by shaking with alkaline $NaHCO_3$ and filtering. The 0.5 (M) sodium bicarbonate adjusted to pH 8.5 actually controls the ionic activity of calcium, through the solubility product of calcium carbonate, during the extraction of calcareous soils. As the carbonate activity in the soil, is raised by this solution, the calcium activity is decreased. Thus some phosphate from the surface of calcium phosphate is extracted through the solubility product of calcium phosphate. As calcium activity decreases, phosphate activity increases, the importance of buffering carbonate during extraction is illustrated by the two trends produced by carbonic acid in calcareous soils.

- a trend towards increased solubility of calcium phosphate as expected with an acid; and
- a trend towards decreased solubility of calcium phosphate owing to the increased calcium activity as $CaCO_3$ is dissolved by carbonic acid.

The reagent also extracts some phosphate from the surface of iron and aluminium phosphate which are more abundant in acid and neutral soil. By repression of Fe and Al activities, the phosphate activity is increased. According to the solubility product principle, the activity of phosphate ions must rise as $a_{Al}$ and $a_{Fe}$ decrease in the presence of $AlPO_4$ and $FePO_4$ ($a$ = activity). Also the carbonate ion added in the reagent, by the solubility product of $CaCO_3$, maintains the Ca activity low enough in all soils (acid, neutral or alkaline) to prevent reprecipitation of the liberated phosphate as calcium phosphate.

Olsen's extractant has a minor disadvantage; it tends to dissolve organic matter resulting in coloured extracts which interferes in colorimetric estimation. Thus activated charcoal is used with the soils to adsorb soluble organic matter. The charcoal must be phosphate free totally. Also carbon dioxide bubbles are formed during colour development which must be completely removed by allowing time or else the $CO_2$ bubbles interfere with colorimetry.

## Reactions

*Exchange reaction*

$$\text{Exchange complex ] Phosphate} + HCO_3^- \longrightarrow \text{Exchange complex] } HCO_3 + \text{Phosphate}$$

*Chemical reaction*

- Extraction

$$Ca_3(PO_4)_2 + 6NaHCO_3 \longrightarrow 3Ca(HCO_3)_2 + 2Na_3PO_4$$

$$3Ca(HCO_3)_2 \longrightarrow 3CaCO_3 + 3H_2CO_3$$

$$2Na_3PO_4 + 3H_2CO_3 \longrightarrow 2H_3PO_4 + 3Na_2CO_3$$

$$Ca_3(PO_4)_2 + 6NaHCO_3 \longrightarrow 3CaCO_3 + 2H_3PO_4 + 3Na_2CO_3$$

- Colour development

$$(NH_4)_6\ Mo_7\ O_{24} \cdot 4H_2O + 6HCl \longrightarrow 7H_2MoO_4 + 6NH_4Cl$$

(Ammonium molybdate) (Molybdenic acid)

$$H_3PO_4 + 12H_2MoO_4 \longrightarrow H_3P(Mo_3O_{10})_4 + 12H_2O$$

(Phosphate) (Phosphomolybdate yellow coloured complex)

$$\text{Phosphomolybdate} + SnCl_2 \xrightarrow{\text{reduction}} \text{Reduced phosphomolybdate}$$

(molybdenum-blue complex of approximate composition)

## Reagents

- Olsen's reagent; 0.5 (M) Sodium bicarbonate solution (pH = 8.5); Dissolve 42.0 g $NaHCO_3$ (L.R.) in double distilled water to give one litre of the solution. Adjust the pH to 8.5 with small amounts of 10% NaOH.
- Activated charcoal (free of P) or Dargo G-60; the activated charcoal is likely to contain traces of P which has to be removed by repeated washings with Olsen's reagent followed by warm distilled water. The material should test free of phosphorus when extracted with Olsen's reagent. The $NaHCO_3$ should also be tested from any phosphate contamination.
- Dickman and Bray's chloromolybdic acid reagent having excess of acid for Olsen's method; weigh 15g of ammonium molybdate (AR) and dissolve in 300 mL of warm water (50°C), cool and filter if necessary. To this, add 400 mL of 10N HCl and make up the volume to one litre. Mix thoroughly and store in an amber glass stoppered bottle.
- Stannous chloride solution; prepare as mentioned in Bray's method.

## Procedure

- Accurately weigh 2.5 g of the soil sample in a 100 mL conical flask and to it add 50 mL of Olsen's reagent.
- Add 1 teaspoon of phosphorus free charcoal.
- Shake the suspension for 30 minutes on a platform type shaker.
- Filter the solution through Whatman No. 40 dry filter paper into clean and dry beakers. Perform a blank without soil (if the filtrate is not clear, return it to the conical flask containing the sample, add more charcoal, shake quickly and filter again).
- Estimate by Dickman and Bray's (excess acid) method.

## Colour development

- Pipette 5 mL of the soil extract into a 25 mL volumetric flask.
- To it add 5 mL of Dickman and Bray's chloromolybdic acid reagent having excess of acid (This must be added drop by drop with constant shaking till the effervescence due to $CO_2$ evaluation ceases).

- Wash the neck of the flask with double distilled water until the contents are diluted to about 22 mL. Add 0.25 mL (5 drops) of the 0.1(M) stannous chloride.
- Make the volume up to the mark with double distilled water.
- Measure the intensity of the blue colour spectrophotometrically at 660 mμ after 5 minutes. Determine the concentration of P from the standard curve perform a blank.

***Note:*** If the concentration of P found is above the range of the method, an aliquot of less than 5 mL is taken and additional extraction solution is added to make up a total of 5 mL of $NaHCO_3$ solution, in order to maintain the proper acidity during colour development. The standard curve is also prepared with same quantity of $NaHCO_3$ included.

*Development of molybdenum blue colour* (pH adjustment in the test solution if pH is at appreciable variance from 3). For precise estimation of phosphorus, an aliquot of the P- containing test solution is pipetted in a 25 mL volumetric flask, adjusted to pH 3 with 4(N) $NH_4OH$ or 4(N) HCl using 2 : 4 dinitrophenol as an indicator (0.25% in $H_2O$) which becomes yellow as pH = 3 is approached from the acid side. If with few drops of indicator yellow colouration is obtained acid is added dropwise until colourless. If the indicator gives a colourless solution indicating a solution pH below 3 drop wise alkali is added just until a yellow colour appears and finally this yellow colour is made faint yellow with dropwise additon of the acid solution.

### Standard curve for phosphorus

- Primary phosphate standard; 50 ppm of phosphorus; Dry (AR) grade potassium dihydrogen phosphate ($KH_2PO_4$) in an air oven at 40–50 °C for one hour and cool in a desiccator. Weigh accurately 0.2195 gm of $KH_2PO_4$ and dissolve in about 400 mL double distilled water in a 1 litre volumetric flask. Then add 25 mL of 7(N) $H_2SO_4$ (approx) and make up the volume to 1000 mL with double distilled water. This gives 50 ppm 'P' solution (Addition of $H_2SO_4$ preserves the solution indefinitely but should be stored in soft glass bottle rather than one of Pyrex to minimize contamination with arsenic).
- Prepare 2 ppm standard (secondary) from the prepared 50 ppm by proper dilution (20 mL of 50 ppm stock diluted exactly to 500 mL for preparing 2 ppm standard. These more dilute stock solutions do not keep well even with addition of tolune and hence must be made up frequently.
- From this 2 mL stock prepare different concentrations of P in 25 mL volumetric flask viz. 0.2, 0.4, 0.6, 0.8, 1.0 ppm by pipetting requisite volume of 2 mL stock.
- To these add 5 mL of extracting reagent (Bray's or Olsen's) and the colour is developed by adding 5 mL of chloromolybdic acid reagent for Bray's and Olsen's method and stannous chloride (1 mL or less).

| P solution | Volume of 2mL stock to be pipetted | Volume of chloromolybdic acid + $SnCl_2$ + double distilled water | Total volume |
|---|---|---|---|
| (ppm) | (mL) | (mL) | (mL) |
| 0.0 | 0.0 | 25.0 | 25.0 |
| 0.2 | 2.5 | 22.5 | 25.0 |
| 0.4 | 5.0 | 20.0 | 25.0 |
| 0.6 | 7.5 | 17.5 | 25.0 |
| 0.8 | 10.0 | 15.0 | 25.0 |
| 1.0 | 12.5 | 12.5 | 25.0 |

- Make up the volume with double distilled water and take the readings after 4–5 minutes at 660 mμ wavelength after properly adjusting the blank (0.00 ppm) to 100% transmittance or 0.00% absorbance.

***Note:*** All reagents including the extraction solution and sample processing chemicals must be included in each of the standard solutions and in the blank employed for preparing the calibration curve. The influence of the extraneous ions and the impurities are thus taken into account. A slight colour in the blank does not matter since the spectrophotometer is set with the blank reading, transmittance = 100 or absorbance = 0.000 and the remaining solutions are read relative to this blank. The standard curve is plotted by taking the spectrophotometer readings along the *y*-axis (ordinate) and the concentration of P(ppm) along the *x*-axis (abscissa). A mean line is drawn through the origin.

## Precautions for P estimation

- For a satisfactory phosphorus procedure constant conditions must be maintained in the blank, standard and test solutions. Contamination must be avoided.
- Alkaline washing powders (which often contains) phosphates must not be used for glassware cleaning. All glass wares should be cleaned with chromic acid and thoroughly washed with double distilled water. For final clearing the glassware may be dipped in or rinsed with 6 (N) HCl after apparently clean, then thoroughly washed with double distilled water.
- Double distilled water must be used for all purposes in P estimation.
- The reagents and filter papers should be as free of phosphorus as possible.
- Dust, perspiration, saliva, tobacco ashes contains appreciable amount of phosphate and therefore should be avoided.
- Pyrex glass apparatus, particularly new ones are to be avoided to minimise contamination with arsenic, which interferes in the analysis.
- Molybdenum blue solutions must not be kept in the volumetric flask after completion of

experiment. The volumetric flasks and the spectrophotometer cuvettes must be washed immediately after use. If a deposit of tin does form inside the glassware dissolve it in hot hydrochloric acid.

- The standard curves for phosphorus may varies slightly according to variation in temperature, slight deterioration of $SnCl_2$ solution and perhaps other factors. It is advisable to construct fresh standard curve for each batch of determination.

## Calculations

*Bray's Method*

$$\text{ppm of P in soil} = \text{ppm of P in solution (obtained from standard curve)} \times 50/5 \times 25/5$$
$$= \text{ppm P} \times 50$$
$$\text{P (kg/ha)} = 112 \times \text{ppm P}$$

*Olsen's method*

$$\text{ppm of P in soil} = \text{ppm of P in solution (obtained from standard curve)} \times 50/2.5 \times 25/5$$
$$= \text{ppm P} \times 100$$
$$\text{P (kg/ha)} = 224 \times \text{ppm P}$$

## Basis for calculation

For both Bray and Olsen's method, 5 mL of soil extract was finally made up to 25 mL and the readings taken in spectrophotometer. Let the ppm obtained from standard curve corresponding to the spectrophotometer reading be $x$ ppm.

Hence by definition $10^6$ mL solution contains $x$ gram of P

Therefore, 25 mL solution will contain $x \times 25/10^6$ gram of P

So, 5 mL original extractant will contain $5 \times 25/10^6$ gram of P

Therefore, 50 mL original extractant will contain $\frac{x \times 25}{10^6} \times \frac{50}{5}$

Hence for *Bray's method*

5 g of soil sample contains $\left(\frac{x \times 25}{10^6} \times \frac{50}{5}\right)$ g of P

Therefore, $10^6$ g soil sample contains $\frac{x \times 25}{10^6} \times \frac{50}{5} \times \frac{10^6}{5}$

$$= \frac{x \times 25 \times 50}{5 \times 5}$$

$$= 50x \text{ ppm P} = (112 \times x) \text{ P kg/ha.}$$

Similarly for *Olsen's method*

2.5 g of soil sample contains $\left(\frac{x \times 25}{10^6} \times \frac{50}{5}\right)$ g of P

Therefore, $10^6$ g soil sample contains $\left(\frac{x \times 25}{10^6} \times \frac{50}{5} \times \frac{10^6}{5}\right)$

$$= \frac{x \times 25 \times 50}{5 \times 2.5}$$

$$= 100x \text{ ppm P}$$

$$= (224 \times x) \text{ P kg/ha.}$$

## 3.12 TOTAL POTASSIUM

Total potassium in soils or minerals is best determined by decomposition of the sample by means of HF; followed by flamephotometric determination.

### Reagents

- 48% HF solution
- 60% $HClO_4$
- 6(N) HCl

### Procedure

- Weigh 0.1000 g finely ground sample (0.16 mm sieve) which has been dried at 100°C for 2 hours in a 30 mL platinum crucible.
- Wet the sample with few drops of water and then add 0.5 mL of $HClO_4$ and 5 mL of 48% HF.
- Place the crucible with lid covering nine-tenths of the top on a sand bath at 200° – 250°C and evaporate the acids to dryness.
- Take care that the solution does not boil vigorously or else spattering may occur [$HClO_4$ drives of $F^-$ because $F^-$ interferes with Fe determination. However, organic matter may condense onto the cover and upper sides of the crucible since it is not completely attacked by the $HClO_4$. If a dark colour is noted on the cover and sides indicating organic matter, dispel the organic matter with faint red heat using Meker burner on the sides of the crucible].
- Remove the crucible from the sand bath and cool sufficiently. Add 5 mL, of 6(N) HCl and dilute the suspension to 2/3 of the volume of the crucible with water.
- Cover the crucible and heat in relatively low flame for 5 minutes to dissolve the residue. When the residue is completely dissolved, transfer the solution in the crucible to a 100 mL volumetric flask.
- Cool and make up the volume for flamephotometric estimation of potassium.

*Note:*

- In high titanium samples, some opalescence tend to develop by Ti precipitation, but when the solution is heated for several minutes in hot water bath the Ti goes into solution.
- Sample of soil, rock, mineral passed through 0.16 mm sieve may be taken from 0.1–1.0 g according to convenience. However, the exact weight taken in the platinum crucible must be noted down.
- The sample is also moistened with few drops of 18 N $H_2SO_4$ instead of water, then 1 mL of $HClO_4$ and 5 mL HF added for the digestion purpose for samples containing less than 2% Ti. Additional portions of $H_2SO_4$, $HClO_4$ and HF acids may be added for digestion purpose, such that a total of 3 treatments and evaporations can be given.
- **Precautions must be taken properly in handling platinum crucible.** Platinum utensils are indispensable for accurate chemical analysis but must be used with care for their expensiveness. Platinum is remarkably resistant to a variety of chemicals, although it is soluble to an appreciable extent in the majority of reagents. That may "properly" be used in platinum utensils. Platinum has a melting point of 1700°C but is appreciably volatile at 1000°C and above.

Since, platinum is too soft in pure condition, it is alloyed for stiffening, usually with a fraction of a per cent of irridium. The life of platinum utensils are maintained well by (*a*) keeping them chemically clean, (*b*) avoiding specific reagents in which platinum is soluble or reactive and (*c*) handling them so as to keep them in good mechanical condition.

### Cleaning procedure

- Crucibles should be cleaned singly
- Soaking in hot water and scrubbing
- If not clean then boiling in 6(N) HCl for few minutes
- If not clean the crucible is dried and fused with potassium pyrosulphate ($K_2S_2O_7$), the melt is poured into dry waste sand and the residue dissolved from the crucible in warm 6(N) HCl.
- Crucibles can be further cleaned by digestion with little HF to which 3 drops of $H_2SO_4$ have been added.
- Crucible is warmed for few minutes in 6(N) HCl, rinsed with distilled water and dried in an oven.

### Specific reagents in which platinum is soluble or reactive must be avoided

- Chlorine attacks platinum. Hence platinum utensils must not be digested in aquaregia from which chlorine is liberated. Ferric chloride in presence of HCl must not be used.
- Hydroxides, oxides, peroxides, nitrites and cyanides of alkalies strongly attack platinum.

### Handling of platinum utensils

- The utensils must be handled in such a manner so as to prevent deformation.
- Platinum-tipped or pure nickel tongs are employed in handling the pt-crucible. Brass, nickel plated or iron tongs should never be used.
- A porcelain plate, beaker or asbestos pad is employed to set the platinum utensil on, never a desk top or ring stand.

## 3.13 AMMONIUM ACETATE EXTRACTABLE POTASSIUM

### Principle

The readily exchangeable plus water soluble $K^+$ is determined in the neutral normal ammonium acetate extract of the soil. The $NH_4^+$ ion provides a sharp and quick separation from the exchange sites while other cations bring about a gradual replacement of either more or less amount of potassium which normally increases with the period of contact. Since, $NH_4^+$ holds highly charged layers together just as $K^+$, the release of non-exchangeable $K^+$ to exchangeable form is retarded during $NH_4OAc$ extraction [Ammonium ions undergoes equilibrium fixation in the 2 : 1 layer silicates, particularly in the highly charged vermiculite interlayer spaces, in exactly the same way as $K^+$, by closure of the interlayer space. The ammonium ions thus fixed undergoes only slow exchange and is reluctant to nitrify, Bower (1951). $Na^+$ ions best replaces $NH_4^+$ and $K^+$ from slow exchange position].

Non-exchangeable K also has been found to contribute appreciably towards potassium availability to crops. The commonly used extractant for such purposes involve hot 1(N)HCl and boiling 1(N)$HNO_3$. The procedure for determination involves either prior removal of exchangeable $K^+$ or conditions made sufficiently vigorous to extract both exchangeable (including water soluble) and a portion of non-exchangeable forms from which the former is subtracted.

### Reagents

- Neutral normal ammonium acetate solution; Dilute 60 mL glacial acetic acid (99.5%) and 75 mL concentrated ammonia solution (sp.gr. 0.91, 25% $NH_3$) to one litre. Mix well, cool and adjust the pH to 7.0 with dilute acetic acid or ammonia solution.
- Potassium chloride solution: 1000 ppm stock solution; Dissolve 1.907 g of AR grade potassium chloride (dried at 60°C for 1 hr.) in distilled water and make up the volume to 1 litre.

### Procedure

- Weigh 5 g soil sample in a 25 mL conical flask.
- Add 25 mL of neutral normal ammonium acetate (pH = 7) and shake for 25 minutes.
- Filter immediately through a dry filter paper (Whatman No.1).
- Reject first few mL of the filtrate.
- Determine the potassium concentration in the extract flamephotometrically after necessary setting and calibration of the instrument.

### Standard curve for potassium

- From the mother stock solution (1000 ppm K), prepare 2, 5, 10, 15 and 20 ppm K solutions in 50 mL volumetric flask by proper dilution.
- Adjust the gas and air pressures of the flamephotometer (as per direction given in the operation manual) and set to the appropriate filter.
- Adjust the flamephotometer reading to zero with the blank (ppm) and at 100 for the maximum ppm, say 20 ppm.

- Construct the standard curve by plotting the flamephotometer readings along Y-axis and the different concentrations (ppm) along X-axis.
- Draw a mean line passing through the origin, *i.e.*, (0, 0) co-ordinate.
- Find out the concentration of the unknown sample by fitting in the standard curve.

### Calculations

$$\text{Available } K^+ \text{ (ppm)} = R \times \frac{\text{volume of extract}}{\text{weight of soil taken in g}}$$

where R = ppm of $K^+$ in the extract, obtained from the standard curve.

### Basis of calculation

Let $x$ ppm is the concentration of the test solution obtained from the standard curve.

Suppose $d$ times dilution was made (if the original extract is too much high in $K^+$).

Then extractant concentration = $(x \times d)$ ppm

Now $10^6$ mL solution contains $(x \times d)$ g of $K^+$

$$\therefore \text{ 25 mL solution contains } \left[\frac{(x \times d)}{10^6} \times 25\right] \text{ g of } K^+$$

$$\text{Again 5 g soil sample contains } \left[\frac{(x \times d)}{10^6} \times 25\right] \text{ g of } K^+$$

$$\therefore 10^6 \text{ g soil sample contains } \left[\frac{(x \times d) \times 25}{10^6} \times \frac{10^6}{5}\right] \text{ g of } K^+$$

$$= \left[(x \times d) \times \frac{25}{5}\right] \text{ ppm of } K^+$$

## 3.14 CATION EXCHANGE CAPACITY (CEC)

### Principle

When a sample of soil is placed in a solution of a salt, such as ammonium acetate, ammonium ions are adsorbed by the soil and an equivalent amount of cations is displaced from the soil into the solution. This reaction is termed as 'cation exchange', and the cations displaced from the soil are referred to as 'exchangeable'. The surface-active constituents of soils that have cation exchange properties are collectively termed as 'exchange complex' and consists for the most part of various clay minerals and organic matter. Soil mineral and organic colloidal particles have negative valence charges that holds dissociable cations and are thus called 'colloidal electrolytes. The cation exchange capacity determination involves measuring the total quantity of negative charges per unit weight of the material. Stated otherwise the total amount of exchangeable cations that a soil can retain is designated as the cation exchange capacity and is usually expressed as milliequivalents per 100 g soil or [cmol($p^+$)$kg^{-1}$]. The determination of CEC is of fundamental importance in soil chemistry research. Adsorption, desorption and leaching of fertilizers, thermodynamic study of ion exchange; retention and release of nutrients, agrochemicals, soil pollutants, all depends upon the exchange capacity of the soil; CEC is also found to be an important parameter for soil classification.

The cation exchange capacity is usually measured by leaching the soil or colloid with neutral normal ammonium acetate. Then, the excess salt is removed by washing with 95% ethanol. The ammonium ion ($NH_4^+$) is then determined by steam distillation with magnesium oxide in an alkaline medium. The ammonia evolved is adsorbed into a known quantity of the standard acid containing methyl red indicator and the excess acid back titrated with a standard alkali.

## Reagents

- 1(N) $NH_4OAc$ adjusted to pH = 7; Dilute 60 mL glacial acetic acid (99.5%) and 75 mL concentrated ammonia solution (sp.gr.0.91, 25% $NH_3$) to 1 litre. Mix well, cool and adjust the pH of the solution to 7.0 with dilute acetic acid or ammonia solution. Alternatively, weigh 77.08 g $NH_4OAc$ and dissolve in one litre distilled water and adjust the pH to 7 carefully with dilute acetic acid or ammonia solution.
- Ethanol 60%
- Ammonium chloride (AR)
- Magnesium oxide – carbonate free, freshly ignited (ignite at 650°C for 2 hours and cool in a desiccator over KOH pellets, store in tightly stoppered bottle)
- Standard $H_2SO_4$; 0.1 (N)
- Standard NaOH; 0.1 (N)
- Standard oxalic acid – 0.1 (N)
- Methyl red indicator
- NaOH; 45%
- Silver nitrate solution about 0.1 (M) : Dissolve 8.5 g $AgNO_3$ in 500 mL water. Add 2 mL concentrated $HNO_3$ and mix well.

## Procedure

- Transfer without loss 10 g of air dry soil sample accurately weighed in a 250 mL beaker and add 50 mL of neutral normal ammonium acetate solution.
- Stir occasionally for an hour cover with watch glass and leave overnight.
- Filter the contents through Whatman No. 44 filter paper receiving the filtrate in a 250 mL volumetric flask.
- Transfer the soil completely on to the filter paper and continue to leach the soil with 1(N) $NH_4OAc$ (using 20 mL at a time), allowing the leachate to drain out completely before adding a fresh aliquot.

- Continue the process, until the flask is full to the mark.
- Preserve this for estimation of exchangeable bases ($Na^+$, $K^+$, $Ca^{++}$ and $Mg^{++}$). The recidue left on the filter paper is intended for determination of cation exchange capacity of the soils.
- Wash the recidue left on the filter paper with 60% alcohol to remove excess ammonium acetate. To ensure this add a pinch of solid $NH_4Cl$ to the recidue on the filter paper and wash with alcohol till the filtrate is free from chloride (as tested with silver nitrate solution, the filtrate is perfectly clear when free from chloride). If the washing is to be interrupted such as for the night, attach a rubber tube to the tail of the funnel and pinch it tight with a clip when there is solution above the level of soil in the filter paper *i.e.,* in no case the soil should dry otherwise loss of ammonia may occur.
- Remove the soil with the filter paper into a 800 mL distillation flask and add about 200 mL of water and about 3 g MgO (one spoonful approximately).
- Add few glass beads and little liquid paraffin so as to avoid bumping and frothing during distillation.
- Pour 100 mL of 45% sodium hydroxide and immediately connect the distillation flask to the condenser and distill ammonia in a known excess of 0.1(N) $H_2SO_4$ (say 25 mL) to which a few drops of methyl red indicator is added. (Continue distillation to collect about 150 mL distillate).
- Back titrate the excess of acid with 0.1(N) NaOH. Standardize NaOH versus oxalic acid and $H_2SO_4$ versus standard NaOH. [see standardization technique, Article No. 3.7].
- Perform a blank distillation without the soil on a similar volume of liquid.

### Calculations

CEC is normally expressed in milliequivalents of the cation per 100 g soil. Milliequivalent means the equivalent weight expressed in milligrams. For instance 20 g of $Ca^{2+}$ represents 1 equivalent or 1000 milliequavalent $Ca^{2+}$. Likewise 18 mg $NH_4^+$ would represent 1 milliequivalent (meq) of $NH_4^+$.

Since 1000 mL of 1(N) acid or alkali = 1.0 g equivalent of any cation.

It follows that 1000 mL of 1(N) acid or alkali = 1000 milliequivalents of any cation.

Therefore, 1 mL 1(N) acid or alkali = 1 milliequivalents of any cation.

$$\text{Cation exchange capacity} = \left[(V_1N_1 - V_2N_2) \times \frac{100}{w}\right] \text{cmol kg}^{-1}$$

where $V_1$ = mL of standard acid taken initially for ammonia absorption,

$N_1$ = normality of standard acid,

$V_2$ = mL of standard base used in back titrating of excess acid,

$N_2$ = normality of standard base,

$w$ = weight of sample in g.

### Fundamental basis of calculation

From standardisation of NaOH versus $H_2SO_4$

$$25 \text{ mL of } \left[\frac{N}{10}\right] H_2SO_4 \equiv X \text{ mL } Y(N) \text{ NaOH}$$

[For convenience in calculation carry out standardisation pipetting 25 mL $H_2SO_4$]

$\therefore$ for back titration process

$$25 \text{ mL of } \left[\frac{N}{10}\right] H_2SO_4 \equiv Z \text{ mL } Y(N) \text{ NaOH.}$$

Hence, equivalent Y(N) NaOH consumed by $NH_4^+$ ion is (X – Z) mL.

We know 1 mL 1(N) NaOH ≡ 1 meq.

$\therefore$ (X–Z) mL Y(N) NaOH = [1 × (X–Z) × Y] meq.

Again $w$ g sample contains [(X–Z)Y] meq.

$\therefore$ 100 g sample will contain $[(X-Z)Y \times \frac{100}{w}$ meq.

where,

X = the volume of standard NaOH required for standardisation of 25 mL (N/10) $H_2SO_4$,

Z = the volume of standard NaOH required during back titration,

Y = the normality of NaOH.

### Alternative calculation in terms of $H_2SO_4$ actually consumed

Refer calculation in Article No. 3.8

Standard $H_2SO_4$ consumed by $NH_4$ = (B–T) mL

1 mL 1(N) $H_2SO_4$ = 1 meq.

(B – T) mL = Z(N) $H_2SO_4$ = [1 × (B – T) × Z] meq.

Again $w$ gram sample correspond to [1 × (B – T) × Z] meq.

$\therefore$ 100 g sample correspond to $(B-T) \times Z \times \frac{100}{w}$

where Z is the normality of standard $H_2SO_4$.

***Notes:***

- *Operational Definition of Cation Exchange Capacity*: Until relatively recently, CEC was defined in rather simple terms in elementary texts as "the sum total of the exchangeable cations that a soil can adsorb" (Brady, 1984) or "the capacity of a soil to hold cations" (Jones, 1982), which does not recognize the importance of the procedure used in determining the outcome of the CEC measurement. The negative charges on soil particles are neutralized by an excess of cations and a deficit of anions (negative adsorption or anion repulsion) over that which would be present in the absence of the solid phase. This essentially means that the major part of the negative charge attracts cations for neutralization and the remainder is involved in repelling anions, which is equivalent to attracting cations. Thus CEC should be defined as

$$CEC = M^{X+}_{Excess} + A^{X-}_{Deficit}$$

where $M^{x+}$ and $A^{x-}$ are the cations and anions in the system, expressed on a charge basis (*i.e.,* $cmol_e\ kg^{-1}$ soil for cations retained, $cmol_a\ kg^{-1}$ for anions). Fortunately the magnitude of the anion deficit is quite small in many cases, varying from 5 to 20% of the CEC at an equilibrium electrolyte concentration of 0.1 M to 0.5 to 2% at 0.001 M, and thus can be and often is ignored.

Schollenberger and Dreibelbis (1930) drew attention to the so-called advantages of using a buffered salt such as $CH_3COONH_4$ (ammonium acetate, $NH_4OAc$) at pH 7 and washing with neutral ethanol rather than water to prevent "hydrolysis" of part of the $NH_4^+$ adsorbed (Kelley, 1948; Schollenberger and Simons, 1945). In hindsight, this was a particularly unfortunate selection of extracting solution for acid, highly weathered soils because the buffered salt automatically raises the pH of the soil to 7, greatly inflating the CEC likely to exist in the soil under field conditions. Despite the fact that it is selected as a standard method for the classification of soils (Soil Survey Staff, 1975), the continuing use of this method on acid soils is inappropriate for a clear understanding of their exchange chemistry.

- The exchangeable cation analysis of saline and alkali soils is subject to difficulties not ordinarily encountered with other soils. Saline and alkali soils commonly contain alkaline-earth carbonates and a relatively high concentration of soluble salts. They may also have low permeability to aqueous solution and to alcohol. The method described above is for non-calcareous soils. In soils containing considerable calcium carbonate, saturation with ammonium will be only partial so long as the carbonate is present because of its solubility in the $NH_4OAc$ solution. The soluble salts should not be washed out of the soils prior to extracting the exchangeable cations, because of significant changes that take place as a result of dilution and hydrolysis. The dissolving of salt therefore necessiates independent determinations of soluble cation contents and correction of the exchangeable cation analysis for their presence, while the occurrence of calcium and magnesium carbonates prevents accurate determination of exchangeable calcium and magnesium. Also, the low permeability of many alkali soils renders the conventional leaching techniques time consuming and inconvenient. Although neutral normal ammonium acetate is the salt solution most commonly used for the extraction of exchangeable cations, some saline and alkali soils fix appreciable amounts of ammonium and potassium under moist condition. This fixation does not interfere with the extraction of exchangeable cations but values obtained for cation exchange capacity are low by amounts equal to the quantity of ammonium fixed. Thus using a cation not subject to fixation for CEC determination is necessary for such soils. $NH_4OAc$ method may also give low result if the soil contain predominantly 1:1 type clay minerals (kaolinitic) or much organic matter. Usually, the CEC is determined by measuring the milliequivalents of sodium adsorbed per 100 g of soil upon treating a soil sample with an excess of normal sodium acetate solution adjusted at pH 8.2. The fact that sodium is a prominent cation in most saline and alkali soils also favours its use in the determination of CEC.

- The method of determination of cation exchange capacity must be reported with the result as because the use of different saturating cations may lead to different results due to variation in cationic size, hydration and electric charge affecting the mechanism of exchange. Cation exchange capacity is not necessarily an absolute constant for a particular soil but may have a range of values according to the cation involved in its determination.
- The direct distillation of soil in an alkaline medium may lead to partial breakdown of organic matter thus introducing a possible error with most surface soils. Also if cation exchange capacity is large, the final titration will consume too much titrant if 10 g soil is taken. Hence for direct distillation of heavy clay soils mainly montmorillonite, it is better to take 5 g soil for analysis.
- The object of using neutral alcohol is to remove excess of occluded ammonium acetate since the ammonium complex undergoes slight hydrolysis if water is the leaching agent. Further the ammonium saturated soil is highly dispersed when in contact with water and the fine particles of the soil show a tendency to pass through the filter paper.
- An alternative and better method is to collect the distillate (ammonia) into a 250 mL conical flask, containing known excess (50 mL) of 2% boric acid solution with mixed indicator (bromocresol green and methyl red) and is titrated with standard sulphuric acid (N/10).

### 3.14.1 Cation Exchange Capacity of Soils Containing Calcium Carbonate

In order to determine the exchange capacity when calcium carbonate is present in soils, recource should be had to a reagent in which the calcium carbonate is insoluble, and which contains a cation which is easily analysed after it replaces the exchangeable calcium. The reagent commonly used is 1(N) sodium acetate adjusted to pH 8.2. When the soil is thoroughly mixed with this reagent, the exchangeable cations are replaced by sodium. Excess of sodium acetate is removed by washing with 95% ethanol. This removal is the critical step in the procedure since exchangeable sodium is easily hydrolysed leading to under estimation of the exchange capacity of soils and therefore removal of salt is monitored via electrical conductivity measurement of the washings. Thereafter, the sodium on the exchange sites are displaced by leaching with 1(N) magnesium nitrate solution adjusted to pH 8.6 and sodium determined through flame photometry.

**Reactions**

$$CH_3\ COONa \rightleftharpoons CH_3\ COO^- + Na^+$$

$$\text{Soil X} + Na^+ + CH_3\ COO^- \rightleftharpoons \text{Soil} - Na + CH_3\ COOX$$

$$Mg\ (NO_3)_2 \rightleftharpoons Mg^{2+} + 2NO_3^-$$

$$\text{Soil} - Na + Mg^{2+} \rightleftharpoons \text{Soil} - Mg + 2Na^+$$

## Reagents

- Sodium acetate solution 1.0 (N), pH 8.2.
  Dissolve 82 g anhydrous sodium acetate or 136 g. Sodium acetate trihydrate in about 90 mL water. Adjust the pH to 8.2 with dilute NaOH or dilute acetic acid and dilute to 1000 mL.
- Ethanol 95%
- Magnesium nitrate 1.0 (N) pH 8.6.

## Procedure

- Weigh 5 g soil sample and place in centrifuge tube.
- Add 33 mL NaOAc, to the tube, stopper and shake for 5 minutes.
- Centrifuge the tubes for 10 minutes at about 8000 rpm.
- Decant the supernatant liquid as completely as possible and discard. Repeat this step four times.
- Add about 30 mL ethanol to each tube, stopper and shake for 5 minutes.
- Centrifuge until the supernatant liquid is clear.
- Decant and discard the supernatant liquid.
- Continue washing until the electrical conductivity of the supernatant liquid from the last washing is between 40 and 55 μmhos $cm^{-1}$ (Check the conductivity of each washing starting from the third. Usually four-five washings are sufficient).
- Replace the adsorbed sodium from the sample by extracting with three 30 mL portions of Mg $(NO_3)_2$ solution.
- Dilute to 100 mL and determine the sodium concentration flamephotometrically or by Atomic Absorption Spectrophotometer.

## Calculations

$$\text{CEC of soil } [\text{cmol}(p^+)\text{kg}^{-1}] = c \times \frac{v_1}{1000} \times \frac{100}{w}$$

where $c$ is the Na content of Mg $(NO_3)_2$ extract (meq/L)

$v_1$ = volume of extract (mL),

$w$ = weight of soil sample (g).

## 3.15 ANION EXCHANGE CAPACITY (AEC)

Anion exchange capacity (AEC) is defined as the quantity of phosphate bound at pH 4 or 5.7. Many anions are often involved in anion exchange reactions viz., $PO_4^{-2}$, $SO_4^{-2}$, $NO_3^-$, $Cl^-$ etc. However, phosphate is usually very suitable for AEC estimation. Under low pH and high concentration, anions may be adsorbed and exchanged on soil colloids. The adsorption usually occurs on surfaces having a positive charge, viz. iron and aluminium hydroxides. The anion retention is related to the nature of anions and that of the soil surface together with amphoteric properties of organic colloids as well as iron and aluminium hydroxides. In highly acidic soil conditions phosphorus acid anions are retained directly on the surface of colloidal particles from the soil solution by adsorption phenomena. The mechanism of anion exchange may be illustrated as follows:

- By the addition of a proton ($H^+$ ion) to the -OH group linked to a sesquioxide clay particle (R) *i.e.*, $Al_2O_3$, $Fe_2O_3$, etc.

$$R\text{–}OH + HOH \longrightarrow R\text{–}OH_2OH$$

$$R\text{–}OH + HCl \longrightarrow R\text{–}OH_2Cl$$

- By the addition of a proton to the functional groups of the organic fraction in acid soil

$$R\text{–}COOH + H^+ \longrightarrow R\text{–}COOH_2{}^+ + Cl^- \rightarrow RCOOH_2Cl$$

The study of AEC of soils helps in understanding the retention and release mechanism of important plant nutrient anions, viz. sulphate, phosphate, nitrate particularly in light textured soils of humid tropics.

## Principle

The method involves initial leaching of the soil with a solution of barium chloride-triethanolamine buffered at pH 8.1, followed by calcium saturation. The Ca-saturated soil is equilibrated with standard phosphoric acid solution and the quantity of phosphorus adsorbed is evaluated. From this adsorbed phosphorus plus phosphorus extracted initially the AEC of the soil is calculated using the formula.

AEC (meq./100 g soil) = [(extractable P + adsorbed P)] expressed as meq./100 g soil.

## Reagents

- Calcium chloride solution; Dissolve 50 g $CaCl_2 \cdot 2H_2O$ in 100 mL of distilled water and adjust to pH = 8.0 with saturated $Ca(OH)_2$ solution.
- Triethanolamine solution; Dilute 90 mL of triethanolamine to 100 mL and adjust the pH to 8.1 with HCl. Dilute to 200 mL and mix equal volume of distilled water containing 100 g of $BaCl_2 \cdot 2H_2O$.
- Phosphoric acid solution [0.01 (M) in $H_3PO_4$].
- Bray's (I) Reagent for P extraction [0.025(N) $NH_4F$ in 0.03 (N) HCl].
- Dikman and Bray's reagent for colour development $KH_2PO_4$. stock solution of P for standard curve construction. (see Article No. 3.11).
- Ethanol – 95%

## Procedure

- Weigh 10 g soil and leach with 100 mL of triethanolamine and wash 6 times with 95% ethanol.
- Leach the soil with 100 mL of $CaCl_2$ solution and wash again.

- Dry the calcium saturated soil at 45°C and weigh into a centrifuge tube sufficient to give a CEC of about 0.2 meq.
- Add 20 mL phosphoric acid solution and shake for half an hour and let stand for 24 hours. Again shake for half an hour. Centrifuge and take 1 mL aliquot for P-estimation.
- In a separate soil sample, extract 'P' with Bray's reagent and determine 'P' colorimetrically using chloromolybdic acid reagent.

### Calculations

Weight of soil taken = 10 g

Volume of phosphoric acid solution added = 20 mL

Volume of aliquot taken = 1 mL

Let this 1 mL is made up to V mL and concentration of P from standard curve = C ppm.

$$\text{Hence first dilution} = \frac{20}{10} = 2 \text{ times}$$

$$\text{Second dilution} = V/1 = V \text{ times}$$

$$\text{Total dilution} = 2V \text{ times.}$$

Therefore, concentration of P in solution phase = (2 × C × V) ppm.

$$\text{Thus P adsorbed} = [\text{P added (ppm)} - 2\,C \cdot V] = X \text{ ppm} = \left(\frac{X}{6.2 \times 10}\right) \text{meq./100g}$$

$$\text{Also, extractable P} = Y \text{ ppm} = \left(\frac{Y}{6.2 \times 10}\right) \text{meq./100g.}$$

$$\text{So, AEC (meq/100g soil)} = \left[\left(\frac{X+Y}{62}\right)\right]$$

## 3.16 EXCHANGEABLE BASES

### 3.16.1 Exchangeable Sodium

#### Principle

Sodium is readily excited in a flame producing an intense yellow light, the yellow colour is primarily due to radiation of 589.6 millimicron wavelength popularly known as D-line of sodium. Other less powerful radiations of different wavelength emitted are effectively blocked by a suitable yellow glass (Na-filter) allowing only the D-line emission to pass through. Thus, if a solution containing sodium ions is fed as a fine spray into a flame under controlled and standard instrumental conditions and the emitted light is passed through a Na-filter, the intensity of the D-line emission can be easily measured photoelectrically and related to the concentration of the sodium in the test solution. The flamephotometer is calibrated with a series of standard sodium chloride solution and then used to determine the unknown sodium concentration of the solution under analysis within the same range.

## Procedure

- Analyze directly the ammonium acetate extract for $Na^+$ and $K^+$ in the flamephotometer.

## Standard Curve for Sodium

- Dissolve accurately weighed 2.542 g NaCl in distilled water and make up the volume to one litre. This gives 1000 ppm stock solution of $Na^+$.
- From this prepare 1, 2, 3, 4, 5, 6, 7 and 10 ppm $Na^+$ by proper dilution.
- Adjust the gas and air pressures of the flamephotometer as per direction given in operation manual and set to appropriate filter.
- Adjust the flamephotometer reading to zero with blank (0 ppm) and 100 for the maximum (100 ppm).
- Construct the standard curve by plotting the flamephotometer reading along X-axis and concentrations along Y-axis.
- Draw a mean line passing through the origin.
- From this graph obtain the sodium concentration of the sample under analysis in milliequivalents per litre or in ppm. If there is a dilution of the original sample, multiply by the dilution factor.
- Check the performance of the flamephotometer at frequent intervals by spraying some standard solutions and adjusting the sensitivity as necessary.

## Calculations

$$\text{Exchangeable } Na^+ \text{ (ppm)} = R \times \frac{\text{volume of extractant}}{\text{weight of soil}}$$

where R = ppm of Na in the extract as obtained from the standard curve. R must include any dilution factor, if used.

also, 1 meq./l Na = 23 ppm Na.

*Notes*

- Standard solutions for Na are also prepared in meq./L rather than ppm. For this a stock solution of 0.05 (N) NaCl is prepared by dissolving accurately weighed 1.4625 g of dry NaCl in 500 mL distilled water. From this stock 2, 4, 6, 8 and 10 mL solution is diluted to one litre respectively to get working standards containing 0.1, 0.2, 0.3, 0.4 and 0.5 milliequivalents per litre.
- In analysis of water or soil extracts, the only ion which may cause serious interference to sodium measurements is calcium. This usually happens when calcium occurs in a much higher concentration than sodium. For instance, water extracts of gypsiferrous soils can contain up to 30–32 meq/L of calcium while sodium level may be less than 1 meq/L. In general for a particular instrument and type of flame, there may be interference effects from some of the other cations or anions present in the test solution and these effects must either be suppressed or measured. The calcium ion sometimes tend to enhance sodium emission and the effect may be measured for a range of calcium : sodium ratios and appropriate corrections applied or in some cases the interference may be suppressed by addition of aluminium nitrates.

Usually, a series of standard NaCl solution is prepared containing 0.1–0.5 meq/L $Na^+$ by dilution from 0.05(N) NaCl solution using saturated calcium sulphate solution in place of water. Since mostly the calcium in soil extract is associated with sulphate these solutions contains about 30 meq/L calcium as sulphate. Next to this, the flamephotometer is calibrated with pure NaCl standard and subsequently the standard containing the calcium sulphate is sprayed and any interference effect due to calcium is measured.

However, sodium may be determined also by atomic absorption spectrophotometry using emission mode. Calcium does not normally interfere in this technique.

### 3.16.2 Exchangeable Calcium and Magnesium

Where atomic absorption spectrophotometry is possible the ammonium acetate extracts can be directly analysed for Ca and Mg. The spectrophotometric standards are prepared in the ammonium acetate solution and both the standard and extracts are read against ammonium acetate as blank. If AAS is not possible the calcium and magnesium are analysed by complexometric titrations using ethylene diamine tetra acetic acid (EDTA).

#### Principle

The method makes use of excellent chelating properties of disodium ethylene diamine tetraacetate (versenate) which forms soluble complexes with metal cations.

The structure of EDTA may be represented as follows:

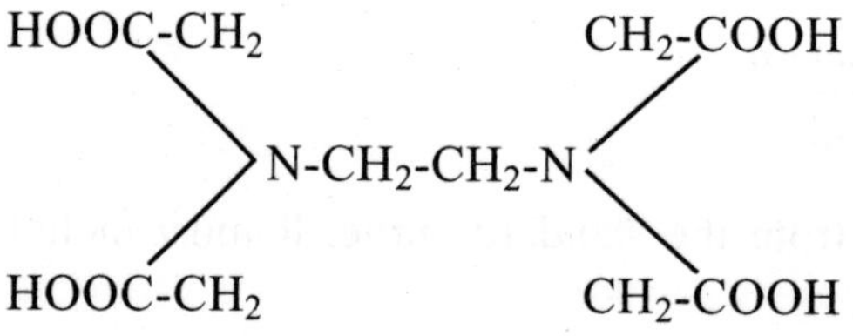


I

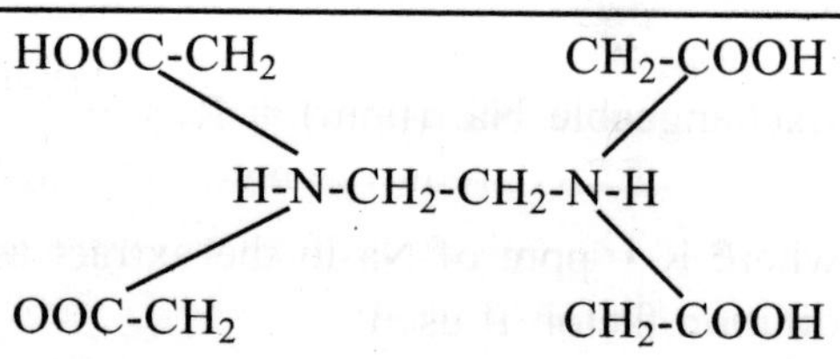


II

The formula II is preferred over I, since it has been shown from measurements of the dissociation constants that two hydrogen atoms are probably held in the form of zwitter ions. For the purpose of simplicity we shall assign the formula $H_4Y$ to EDTA : the di-sodium salt is therefore $Na_2H_2Y$ and affords the complex-forming ion $H_2Y_2$ in aqueous solution; it reacts with all metals in a 1:1 ratio. The reactions with cations $M^{2+}$, can be written in the general form as:

$$M^{2+} + H_2Y^{2-} = MY^{2-} + 2H^+$$

$$M^{3+} + H_2Y^{2-} = MY + 2H^+$$

or

$$M^{n+} + H_2Y^{2-} = (MY)^{(n-4)+} + 2H^+$$

One mole of complex forming $H_2Y^{2-}$ reacts with one mole of the metal ion and in each case two moles of hydrogen are formed. It is evident from the equations above that dissociation of the complex will be governed by the pH of the solution; lowering the pH will decrease the stability of metal-EDTA complex. The more stable the complex, the lower the pH at which an EDTA titration of the metal ion in question is carried out.

The stability of a complex is characterised by the stability constant (or formation constant) K:

$$M^{n+} + Y^{4-} = (MY)^{(n-4)+}$$

$$\therefore \qquad K = \frac{[(MY)^{(n-4)+}]}{[M^{n+}][Y^{4-}]}$$

Assuming the fully ionised form of EDTA *i.e.,* the ion $Y^{4-}$ has been taken into account, but at two pH-values the species $HY^{3-}$, $H_2Y^{2-}$, $H_3Y^-$ and even undissociated $H_4Y$ may be present; stated otherwise only a part of the EDTA uncombined with metal may be presented as $Y^{4-}$. Further, the metal $M^{n+}$ is assumed to be uncomplexed *i.e.,* in aqueous solution, it is simply present as the hydrated ion.

The success of an EDTA-titration depends upon the precise determination of the end point. The most common technique is to use metal-ion indicators. The requisites of a metal ion indicator for use in the visual detection of end point include:

- The colour reaction must be such that before the end-point when nearly all the metal ion is complexed with EDTA, the solution is strongly coloured.
- The colour reaction should be specific.
- The metal-indicator-complex must posses sufficient stability, otherwise because of dissociation, a sharp colour change is not obtained. The metal indicator complex however, must be less stable than the metal-EDTA complex to ensure that, at the end point, EDTA removes metal ions from the metal-indicator complex.
- The colour contrast between the free indicator and the metal-indicator complex should be such as to be readily observed.

The use of a metal ion indicator in an EDTA titration may be written as:

$$\text{M-In} + \text{EDTA} \longrightarrow \text{M-EDTA} + \text{In}$$

The reaction will proceed if the metal-indicator complex (M-In) is less stable than the metal-EDTA complex (M-EDTA). The former dissociates to a limited extent and during the titration the free metal ions are progressively complexed by the EDTA until ultimately the metal is displaced from the complex (M-In) to leave the free indicator (In).

Some of the metal ion indicators used for calcium and magnesium are discussed below:

*Solochrome dark blue or calcon*

This is sodium 1 – (2- hydroxy – 1 – napthylazo) – 2 – napthol – 4 – sulphonate

An important application of this indicator is in the complexometric titration of calcium in presence of magnesium. This must be carried out at a pH of about 12.3 in order to avoid interference of magnesium (obtained with a diethylamine buffer; 5cc/100cc solution). Under such condition magnesium is precipitated quantitatively as magnesium hydroxide. The colour change at end point is from pink to pure blue.

*Patton and Reeder's indicator*

The indicator is 2-hydroxy-1-(2-hydroxy-4-sulpho-1-napthylazo)-3-napthoic acid commonly abbreviated as HHSNNA. Its main application is direct titration of calcium, particularly in presence of magnesium. A sharp colour change is from wine red to pure blue is obtained when calcium ions are titrated with EDTA at pH values between 12 and 14.

*Murexide*

This is ammonium salt of purpuric acid.

Murexide is of interest because it was probably the first metal ion indicator to be employed in the EDTA titration. The murexide may be employed for direct EDTA titration of calcium at pH = 11; the colour change at end point is from red to blue violet.

## Fundamental concept of titration

When calcium ions are titrated with EDTA a relatively stable calcium complex is formed

$$Ca^{2+} + H_2Y^{2-} \longrightarrow CaY^{2-} + 2H^+$$

With magnesium ions, a somewhat less stable complex is formed

$$Mg^{2+} + H_2Y^{2-} \longrightarrow MgY^{2-} + 2H^+$$

The magnesium-indicator complex is more stable than the calcium-indicator complex but less stable than Magnesium-EDTA complex.

*i.e.,* $$[\text{Mg-EDTA}] < [\text{Mg-In}] > [\text{Ca-In}]$$

Consequently, during titration of solution containing magnesium and calcium ions with EDTA in presence of Eriochrome Black T the EDTA first reacts with the free calcium ions, then with the free magnesium ions, and finally with the magnesium-indicator complex. Since magnesium-indicator complex is wine red in colour and the free indicator is blue between pH = 7–11, the colour of the solution changes from wine red to blue at the end-point .

$$\underset{\text{(red)}}{[\text{Mg-EBT}^-]} + H_2Y^{2-} = MgY^{2-} + \underset{\text{(blue)}}{HEBT^{2-}} + H^+$$

The titration with EDTA, using Solochrome Black (Eriochrome Black T) as indicator gives the total calcium plus magnesium content. To determine the individual elements, calcium may be estimated by titration using Patton and Reeder's indicator. The difference between the two gives the estimate of magnesium.

Traces of many metals interfere in the determination of calcium and magnesium using Eriochrome Black T indicator (e.g., Co, Ni, Cu, Zn, Hg and Mn). The interference can be overcome by addition of a little hydroxylamine hydrochloride which reduces some of the metals to their lower valency states and also of sodium or potassium cyanide complexes. Iron may also be rendered harmless by addition of a little sodium sulphide.

## Reagents

- Hydrochloric acid (Analytical grade)
- Nitric acid (Analytical grade)

- Standard Ca solution (0.0 1N) primary standard; Dissolve 0.2502 g of pure calcium carbonate (analytical grade, dried at 110°C overnight) with minimum quantity of concentrated HCl dropwise (10 mL of 3N HCl may be used). Warm the solution to expel $CO_2$ and then dilute to 500 mL in volumetric flask.
- EDTA solution approx. 0.01 (N); Dissolve 2.0 g disodium dihydrogen ethylenediamine tetraacetate ($Na_2\ H_2\ C_{10}\ H_{12}O_8\ N_2 \cdot 2H_2O$) and 0.05 g magnesium chloride hexahydrate in one litre water and standardize against standard calcium solution.
- Sodium hydroxide solution—10%
- $NH_4Cl - NH_4OH$ Buffer (pH = 10); Dissolve 67.5 g ammonium chloride (AR) in 570 mL of concentrated ammonia (sp.gr.0.91) and dilute to one litre.
- Eriochrome Black T indicator; Take 100 mL of ethanol and dissolve 4.5 g hydroxylamine hydrochloride in it. Now add 0.5 g of the indicator and prepare solution.
- Calcon indicator; Dissolve 0.20 g of the dyestuff (calcon) in 50 cc methanol; (ethanol may also be used). Prepare fresh solution weekly.

***Note:*** Sodium diethyl dithio carbamate crystals or 2% sodium cyanide solution, (used generally to eliminate interference arising due to presence of Cu, Zn, Fe, Mn, Sn, if present in appreciable amounts. However, in irrigation water and water extract of soil interfering ions are negligible and can be neglected).

## Procedure

### Pretreatment of $NH_4OAc$ extract

- Ammonium acetate may interfere in EDTA – tiltration and is therefore destroyed by oxidation with a mixture of HCl and $HNO_3$.
- Pipette 100 mL of $NH_4OAc$ extract into a 500 mL beaker and evaporate carefuly to dryness, cool, and add 5 mL concentrated HCl(washing down salts on wall of the beaker), followed by 1 mL $HNO_3$.
- Cover with a watch glass immediately.
- When vigorous reaction has ceased evaporate the solution to dryness in a fume hood. Cool and add 1 mL concentrated HCl followed by 20 mL water, stir well and filter the solution through a Whatman No. 40 filter paper into a 100 mL volumetric flask.
- Rinse the beaker 3–4 times, collecting the rinsings through the filter paper into the volumetric flask.
- Make up the volume with distilled water.
- Prepare a blank solution by taking 100 mL ammonium acetate solution by the same procedure.

### Calcium

- Pipette 5 mL of the solution (or a suitable aliquot) into a 100 mL conical flask or into a porcelain dish and dilute it approximately to 25 mL (add 20 mL water, if 5 mL aliquot is taken).

- Add 1 mL or more of 10% NaOH to raise the pH to 12. (Check the pH with a pH⁻ meter, if necessary), (2–3 crystal of carbamate may be added if required).
- Add 10–12 drops of calcon indicator stir the solution and titrate with standard EDTA solution until the colour changes from pink to pure blue.
- To check the end point accurately, perform a blank titration taking 25 mL water instead of sample solution and adding other reagents in the similar manner.
- Also determine the blank correction by titrating a similar aliquot of blank solution.

## Calcium plus Magnesium

- Pipette 5 mL of the solution or a suitable aliquot (containing not more than 0.1 meq. $Ca^+$ Mg) into a dish or 100 mL conical flask and dilute to about 25 mL.
- Add 10 mL of $NH_4Cl$-$NH_4OH$ buffer, (2–3 carbamate crystals if required) and 3–4 drops of Eriochrome Black T indicator.
- Titrate against standard EDTA solution until colour changes from red to permanent blue colour.
- Perform a blank by replacing sample with 25 mL distilled water.
- Also determine the blank correction.

Standardise the EDTA solution with standard Ca solution using Eriochrome black-T indicator or calcon indicator (see note below), using the procedure for calcium-estimation.

***Note:***

- For analysing water samples, $NH_4OAc$-pretreatment is not required and hence 'blank correction' is also not done.
- If the EDTA solution is prepared by dissolving 2.0 g of the salt in one litre water without addition of 0.05 g magnesium chloride hexahydrate, then use calcon indicator during standardization of EDTA with standard calcium solution.

## Calculations

Exchangeable calcium (meq/100 g soil).

$$= \left[\frac{V_1 - V_2}{V_3} \times V_4 \times N \times \frac{100}{w}\right]$$

where $V_1$ = volume of EDTA required for sample aliquot titration (calcon), mL

$V_2$ = volume of EDTA required for blank titration (calcon), mL

$V_3$ = volume of aliquot, mL

$V_4$ = total volume of original $NH_4OAc$ extract, mL

N = normality of EDTA,

$w$ = weight of sample in g.

Exchangeable (Ca + Mg), [meq./100 g soil]

$$= \left[\frac{V_5 - V_6}{V_7} \times V_4 \times N \times \frac{100}{w}\right]$$

where $V_5$ = volume of EDTA (mL) required for sample aliquot titration using EBT,

$V_6$ = volume of EDTA (mL) required for blank aliquot titration using EBT,

$V_7$ = volume of aliquot taken (mL),

$V_4$ = total volume of original $NH_4OAc$ extract (ml),

N = normality of EDTA,

$w$ = weight of sample taken in g.

*Note:* 1 mL 0.01 (N) EDTA ≡ 0.2004 mg $Ca^{2+}$ ≡ 0.1216 mg $Mg^{2+}$

## 3.17 EXCHANGEABLE CALCIUM AND MAGNESIUM IN CALCAREOUS SOILS

A KCl solution buffered at pH = 8.3 by triethanol amine is used as an extractant.

### Reagents

- KCl (1.0 N) – triethanolamine buffer solution; Dissolve 74.6 g potasium chloride in about 500 mL water, add 25 mL triethanolamine (sp.gr.1.12) and stir well. Dilute to about 850 mL and mix. Adjust the pH to 8.2 with 1.0 (N) HCl. About 85 mL 1(N) HCl will be required. Dilute to one litre.

### Procedure

- Weigh 10 g air dry sample into a 100 mL beaker and add 40 mL KCl-triethanolamine solution. Stir thoroughly frequently for 20 minutes.
- Filter the suspension through Whatman No. 40 filter paper into a 100 mL volumetric flask. Leach with 20 mL portions of the buffer solution to bring the volume of leachate to 100 mL.
- The extract can be analyzed for Ca and Mg directly by AAS. Prepare the standards (in terms of meq/L) in the KCl-triethanolamine solution.
- Read standards and test samples against the buffer solution as blank.
- If AAS is not possible, determine Ca and Mg by EDTA/versenate method as described previously. (No pretreatment with HCl-$HNO_3$ is required).
- Determine a blank for KCl-triethanolamine solution using Eriochrome black T and calcon indicators.

### Calculations

Exchangeable Ca (meq/100 g soil)

$$= \left[\frac{V_1 - V_2}{V_3} \times N \times \frac{V_4}{w} \times 100\right]$$

where $V_1$ = volume of EDTA for sample titration using calcon (mL),

$V_2$ = volume of EDTA for blank titration using calcon (mL),

$V_3$ = volume of aliquot taken (mL),

$V_4$ = total volume of KCl-TEA extract (mL),

N = normality of EDTA,

$w$ = weight of sample in g.

Exchangeable (Ca + Mg) in meq./100 g soil

$$= \left[\frac{V_5 - V_6}{V_7} \times N \times \frac{V_4}{w} \times 100\right]$$

where $V_5$ = volume of EDTA required for sample titration using EBT(mL),
$V_6$ = volume of EDTA required for blank titration using EBT (mL),
$V_7$ = volume of aliquot taken (mL),
$V_4$ = total volume of KCl-TEA extract (mL),
N = normality of EDTA,
$w$ = weight of sample in g.

## 3.18 MICRONUTRIENTS: DTPA EXTRACTABLE $ZN^{2+}$, $CU^{2+}$, $FE^{2+}$, $MN^{2+}$ (LINDSAY & NORVELL,1978)

### Principle

The micronutrient cations can be estimated in a single extraction with diethylene triamine pentaacetic acid (DTPA) which has excellent chelating property with the micronutrient elements. Adequate precaution must be taken against any likely contamination from the reagents and glass wares in micronutrient assay work. Only double distilled water should be used. Specific hallow cathode lamps for each elements are used on AAS and requisite standards for instrument calibration are prepared as per instructions in the operation manual.

### Reagents

- DTPA – 0.005 M solution; Weigh 1.967 g of DTPA and 1.470 g $CaCl_2 \cdot 2H_2O$ in a beaker. To this add 20–25 mL of double distilled water and 13.3 mL of Triethanolamine (TEA) followed by 100 mL of double distilled water. Transfer to one litre volumetric flask with 3–4 washings and make up the volume up to the mark with double distilled water. Adjust the pH to 7.3 with dilute HCl (1:5).
- TEA (Triethanolamine)
- $CaCl_2 \cdot 2H_2O$ (AR)
- Dilute HCl (1 : 5)

### Procedure

- Weigh accurately 10 g of soil sample in a 100 mL conical flask and add 20 mL of DTPA extractant, shake for 2 hours.
- Filter the extract through Whatman No. 42 filter paper.
- Estimate the micronutrient cations ($Zn^{2+}$, $Cu^{2+}$, $Fe^{2+}$, $Mn^{2+}$) with the help of atomic absorption spectrophotometer.

### Calculation

Micronutrient elements (ppm) = 2 × concentration (ppm) obtained from AAS × dilution factor (if any).

***Note:*** The standard atomic conditions for the respective micronutrient elements are to be followed from Instruction Manual of the Atomic Absorption Spectrophotometer.

## 3.19 DETERMINATION OF AVAILABLE BORON

### Principle

Boron exists in organic as well as in inorganic forms in soils, and most of the water soluble B-compounds are highly available. Boron deficiency is encountered in calcareous and acid soil, whereas toxicity occurs in salt-affected soils as well as in soils irrigated with high B water. Gupta (1967) developed a rapid and easy method wherein the soil could be extracted by boiling with water directly on a hot plate. Use of azomethine-H (John *et al.*, 1975) in place of carmine or curcumin has further simplified the determination of hot-water soluble B. Azomethine-H forms a stable coloured complex with $H_3BO_3$ at pH 5.1 in aqueous media, which retains proportional absorbance-concentration properties for several hours independent of the presence of a variety of salts.

### Hot-water soluble boron (Gupta, 1967)

### Reagents

- **Buffer solution:** Dissolve 250 g of ammonium acetate and 15 g of EDTA (disodium salt) in 400 mL of distilled water, slowly add 125 mL of glacial acetic acid and mix thoroughly.
- **Azomethine-H reagent:** Dissolve 0.45 g of azomethine-H in 100 mL of 1% L-ascorbic acid solution. Store in polypropylene bottle in a refrigerator. Prepare fresh solution every week.
- **Boron standard solution:** Dissolve 0.114 g of AR grade boric acid ($H_3BO_3$) in distilled water and make the volume to 100 mL. Each mL of this solution contains 20 μg of B. Dilute 0.5, 1, 2, 3, 4, 5, 10, 20, 30, 40 and 50 mL of this stock solution to 100 mL with distilled water to have solution concentrations of 0.1, 0.2, 0.4, 0.6, 0.8, 1.0, 2.0, 4.0, 6.0, 8.0 and 10 μg B $mL^{-1}$, respectively.
- Activated charcoal.

### Procedure

- Weigh 20 g of air-dry (20-mesh) soil sample in a 250 mL quartz or other boron-free conical flask and add 40 mL of distilled water.
- Add 0.5 g of activated charcoal and boil for 5 minutes on a hot plate, filter immediately through Whatman No. 42 filter paper.
- Cool the contents to room temperature and transfer 1 mL aliquot of blank (filtrate obtained after boiling of distilled water and activated charcoal), diluted B standard or sample filtrate into 10 or 15 mL polypropylene tubes.
- Add 2 mL of buffer solution and mix.
- Add 2 mL of azomethine-H reagent, mix and after 30 minutes read the absorbance at 420 nm on a spectrophotometer.

- Prepare a standard curve plotting B concentrations (0 to 10 μg B $mL^{-1}$) on X-axis and absorbance on Y-axis.
- Refer the absorbance readings of sample aliquots to the standard curve to obtained B content of aliquots.

### Calculation

Available B (mg $kg^{-1}$ or ppm) in soil = A × 2

where, A is the Boron content (μg) obtained from standard curve.

### Precautions

- The use of borosilicate glass for B determination and storage of reagents should be avoided as far as possible.
- Fresh solution of azomethine-H reagent every week should be prepared.
- Tetrasodium salt of EDTA to be used instead of disodium salt for preparation of buffer solution in order to eliminate interferences due to Al in acid soils dominated by high exchangeable aluminium.

## 3.20 DETERMINATION OF AVAILABLE MOLYBDENUM

### Principle

Amongst the essential plant nutrients, Mo is required in the smallest quantity and is absorbed by plants in the form $MoO_4^{2-}$ ions. Availability of Mo, unlike other elements, is high in the alkaline pH range. As it is required in very small amount, tests for availability of Mo are rarely carried out, particularly in alkaline soils. Ammonium oxalate method is mostly used for extraction of available Mo from soil. The extractant is advantageous since it is easy to prepare, does not allow a significant change in pH during shaking due to buffering action, and molybdenum forms stable complex with oxalic acid. The extraction is based on the replacement of $MoO_4^{2-}$ ions from exchange sites, which is irreversible due to formation of strong Mo-oxalic acid complexes.

### Reagents

- 50% potassium iodide solution: Dissolve 50 g of KI in 100 mL of double distilled water.
- 50% ascorbic acid solution: Dissolve 50 g of ascorbic acid in 100 mL of double distilled water.
- 10% sodium hydroxide solution: Dissolve 10 g of NaOH in 100 mL of double distilled water.
- 10% thiourea solution: Dissolve 10 g thiourea in 100 mL of double distilled water and filter. Prepare fresh solution before use.
- Toluene-3,4-dithiol solution (commonly called dithiol): Weigh 1.0 g of AR grade melted dithiol in a 250 mL glass beaker. Add 100 mL of the 10% NaOH solution and warm the contents up to 51°C with frequent stirring for 15 minutes. Add 1.8 mL of thioglycolic acid and store in a refrigerator.
- 10% tartaric acid: Dissolve 10 g in 100 mL of double distilled water.

- Iso-amyl acetate
- *Ethyl alcohol*
- Ferrous ammonium sulphate solution: Dissolve 63 g of the salt in about 500 mL of double distilled water and then make the volume to one litre.
- Extracting reagent: Dissolve 24.9 g of AR grade ammonium oxalate and 12.6 g oxalic acid in water and make the volume to one litre. Adjust the pH at 3.3 with dil. HCL/ dil. ammonia solution.
- Standard stock solution (100 μg Mo $mL^{-1}$): Dissolve 0.150 g of AR grade $MoO_3$ in 100 mL of 0.1 N NaOH, make slightly acidic with dil. HCL and make the volume to 1L.
- Working standard solution (1 μg Mo $mL^{-1}$): Dilute 10 mL of the stock solution of 1L.

## Procedure

- Weigh 25 g of air-dry soil sample in 500 mL conical flask.
- Add 250 mL of the extracting solution and shake for 10 hours.
- Filter through Whatman No. 50 filter paper. Transfer 200 mL of the clear filtrate in a 250 mL glass beaker and evaporate to dryness on a water bath.
- Heat the contents in the beaker at 500ºC in a furnace for 5 hours to destroy organic matter and oxalates. Keep overnight.
- Digest the contents with 5 mL of $HNO_3$ – $HClO_4$ mixture (4 : 1), then with 10 mL of 4 N $H_2SO_4$ and $H_2O_2$ each time bringing to dryness.
- Add 10 mL of 0.1 N HCl and filter. Wash the filter paper with additional 10 mL of 0.1 N HCl and double distilled water until the volume of the filtrate becomes 100 mL.
- Run a blank side by side (without soil).
- Take 50 mL of filtrate in 250 mL separatory funnels and add 0.25 mL of the ferrous ammonium sulphate solution and 20 mL of double distilled water and shake vigorously.
- Add excess of KI solution and clear the liberated iodine by adding ascorbic acid drop by drop while shaking vigorously.
- Add 1 mL of tartaric acid and 2 mL of thiourea solution and shake vigorously.
- Add 5 drops of dithiol solution and allow the mixture to stand for 30 minutes.
- Add 10 mL of iso-amyl acetate and separate out the contents (green colour) in colorimeter tubes/cuvettes.
- Read the colour intensity at 680 nm (red filter).

*Preparation of standard curve for Mo*

- Measure 0, 2, 5, 10, 15 and 20 mL of the working standard Mo solution containing 1 μg Mo $mL^{-1}$ in a series of 250 mL separatory funnels.
- Proceed for colour development as described above for sample aliquots.
- Read the colour intensity and prepare the standard curve by plotting Mo concentrations against readings.

### Calculation

$$\text{Available Mo in soil } (\mu g\ g^{-1}) = A \times \frac{250}{200} \times \frac{100}{50} \times \frac{1}{25} = \frac{A}{10}$$

where A stands for Mo concentration in μg as obtained on X-axis against a sample reading.

## 3.21 ESTIMATION OF SOIL POLLUTANTS (Cd, Cr, Ni, and Pb)

### Theoretical Background

Pollution is the change in characteristics of air, water, soils and food that can adversely affect the health, survival or activities of humans or other living organisms (Miller 1991). Holdgate (1979) defined pollution as "the introduction by man into the environment of substances or energy liable to cause hazards to human health, harm to living resources and ecological systems, damage to structures or amenity or interferences with the legitimate uses of environment."

Soil pollution by different sources from across several parts of India have been reported by Rattan *et al.* (2002). Soil contamination by heavy metals (viz. lead, cadmium, chromium, mercury, nickel, selenium, arsenic and so on) due to disposal of sewage and industrial effluents is a matter of serious concern since the same can enter the food chain and result in health hazard. Furthermore, in India selenium toxicity in soils/seleniferous soils (soils containing more than 0.5 mgSe/kg), plants as well as in groundwater have been identified in Hoshiarpur and Nawanshahar districts of Indian Punjab which may cause chronic selenosis in animals and human mainly due to consumption of forages and grains produced on seleniferous soils (Sanyal and Dhillon, 2006).

The soil resource (a vital environmental segment) is recognized as one of the most valuable life-supporting basic natural resource on whose proper use depends socio-economic development of any country. Intensive agriculture without any serious attention to our soil health (Quality) has degraded our soils beyond limit. Amongst several known factors soil contamination by toxic elements is posing severe threat both for the environment as well as human health. Although soil can act as a physical filter by its sieving action, a chemical filter by adsorbing and precipitation of chemical substances and biological filter by decomposing of organic materials, it does not have infinite capacity to perform such functions. Any addition to soil of substances that may exert adverse effect on its functioning is known as soil contamination. Since most soil possesses buffering capacity, it usually takes some time for the negative effects to become visible and then the soil can be considered as a polluted one and the process as soil pollution.

Soils in many parts of the world have become polluted with Cd, Cr, Ni and Pb, among other metals, due to mining and smelting activities, disposal of sewage sludge and other wastes. The range of Cr, Ni and Pb in soils normally varies from 10 to 100 mg/kg and that is of Cd below 1 mg/kg. Cadmium and lead pose considerable health hazard to humans and animals as environmental pollutants. Chromium is at higher concentration, a toxic element for mammals and other animals, while Ni is a possible carcinogen for abscessing metal pollution in soil from plant availability point of view, variety of extraction procedures are available. The choice of procedure to be used will depend on the metals of interest and the objective of determination.

### 3.21.1 Total Sorbed Cadmium, Chromium, Nickel, and Lead

The U.S. Environmental Protection Agency (1986) has adopted an acid digestion procedure (EPA method 3050) for total sorbed metals viz., Cadmium, Chromium, Nickel, and Lead. This method 3050 gives a reliable measure of the amounts of metals added to soils and nonsilicates from industrial sources. The method is discussed below:

#### Reagents

- Nitric acid: concentrated and 1:1 dilution in deionized water.
- 30% hydrogen peroxide.
- Concentrated Hydrochloric acid.

#### Method

- Weigh 2 g of air-dried soil (< 1 mm) in a 150-mL beaker and add 10 mL of 1:1 $HNO_3$.
- Place the sample on a hot plate, cover with a watch glass, and heat (reflux) at 95°C for 15 min.
- Cool the digest and add 5 mL of conc. $HNO_3$. Reflux for an additional 30 min at 95°C.
- Repeat the last step, and reduce the solution to about 5 mL without boiling (by only partially covering the beaker).
- Cool the sample again and add 2 mL of deionized water and 3 mL of 30% $H_2O_2$.
- Heat the sample gently to start the peroxide reaction keeping the beaker covered. If effervescence becomes excessively vigorous, remove the sample from the hot plate. Continue to add 30% $H_2O_2$ in 1-mL increments, followed by gentle heating until the effervescence subsides.
- Add 5 mL of conc. HCl and 10 mL of deionized water and reflux the sample for an additional 15 min without boiling.
- Cool and filter the sample through a Whatman No. 42 filter paper. Dilute to 50 mL with deionized water. Analyze for Cd, Cr, Ni, and Pb by Atomic Absorption Spectrophotometry (AAS) or Inductively Coupled Plasma (ICP) spectroscopy.

### 3.21.2 Available Heavy Metals (Cd, Cr, Ni, and Pb)

Total elemental analysis is not necessary for many environmental and agricultural applications, it to determine the total contents of elements in soil. To quantify the labile or available fraction, extraction with DTPA is a suitable procedure. This method was originally developed to identify inadequate supplies of the plant-available micronutrient metals, Cu, Zn, Fe, and Mn in near-neutral and calcareous soils for which, alkaline buffered 0.005 M DTPA is recommended at 2:1 extractant soil ratio (Lindsay and Norvell,1978). However, for extracting available heavy metals alkaline buffered (pH 7.3) DTPA extractant at 5:1 extractant soil ratio was found to be effective in metal-contaminated acid and calcareous soils.

#### Reagents

DTPA extracting solution:the solution is 5 mM DTPA + 10 mM $CaCl_2$ + 0.1 M triethanolamine (TEA). To prepare 5 L of this solution, dissolve 74.6 g of reagent-grade TEA in approximately

1 L of deionized water (< 10 megaohm-cm resistivity). Next dissolve 9.835 g DTPA and 7.35 g $CaCl_2 \cdot 2H_2O$ in the TEA solution. Allow time for DTPA to dissolve, and dilute to approximately 4 L. Adjust the pH to 7.3 ± 0.05 with 1:1 HCL (approximately 42 mL required) and dilute to 5 L final volume. The solution is stable for several months.

### Method

- Place 5 g of air-dried soil ( > 2 mm) in a 125-mL. Erlenmeyer flask.
- Add 25 mL of the DTPA extracting solution and shake for 2 hour at a speed of 120 cycles $min^{-1}$.
- After shaking filter through a Whatman No. 42 filter paper. Refilter if filtrate is turbid.
- Analyze metals in filtrate by atomic spectroscopy.

***Note:*** The standard solution as well as standard atomic conditions for the respective micronutrient elements are to be followed from instruction manual of the Atomic Absorption Spectrophotometer.

## 3.22 ARSENIC DETERMINATION BY CONVERSION TO THEIR HYDRIDES AND ASPIRATION INTO AN AAS

Arsenic is ubiquitous in nature and is found in detectable concentrations in all environmental matrices. The occurrence of As in the continental crust of Earth is usually given as 1.5 to 2.0 mg/L. The distribution of arsenic in nature is extremely variable, showing little correlation with geological formation, climate, or soil. Numerous minerals, rocks, sediments and soils contain arsenic partly as constituent of sulfide minerals or complex sulfides of metal cations and partly as a constituent retained by soils and/or sediments in occluded or adsorbed forms. The latter is manifested primarily by the adsorption or occlusion of As on hydrous Al and Fe oxides, but these are not necessarily the only source. Arsenic is also adsorbed on clay colloid, is bound to organic matter and may form slightly water soluble compounds with Al, Fe, Ca and Mg in the soil matrix. Some of the more common minerals in soils are arsenopyrite (FeAsS), Orpiment ($As_2S_3$) etc.

Arsenic is a labile element and can exist in several forms and oxidation states (–3, 0,+3 and +5 valence in nature). In strongly reducing environments, elemental As and As(III) can exist but As(V) is the stable oxidation state in aerobic environments.

In reduced environments such as sediments, the methanogenic bacteria reduces As(V) to As(III) and methylates it to methyl arsenic acid. Standard for maximum allowable As concentration in drinking water has been set to be 0.01 mg/L by World Health Organization (WHO). Acute As poisoning in human beings is characterized by central nervous system effect, leading to coma and eventually death. Chronic intoxication results in neurological disorders, muscular weakness, loss of appetite, nausea, and skin disorders such as hyper-pigmentation and keratosis.

### Principle

The system consists of an atomic absorption spectrophotometer and a hydride generator. Most atomic absorption spectrophotometer manufacturers now offer hydride generators or accessories that can quickly be attached to a spectrophotometer and are rather simple to operate. The system involves generating arsine as hydride, transferring the hydride to a quartz cell mounted in the light beam of the spectrophotometer, decomposing the hydride within the confines of the cell heated externally by an air-acetylene flame and finally obtaining a measured absorption signal.

Arseneous acid, the As(III) oxidation state of arsenic are instantaneously converted by sodium borohydride reagent in acid solution to their volatile hydrides. The hydrides are purged continuously by argon or nitrogen into an appropriate atomizer of an atomic absorption spectrophotometer and converted to the gas phase atoms. The sodium borohydride reagent by rapid generation of the elemental hydrides in an appropriate reaction cell, minimizes dilution of the hydrides by the carrier gas and provides rapid, sensitive determination of arsenic. At room temperature and solution pH values of 1 or less, arsenic acid, the As (V) oxidation state of arsenic, is reduced relatively slowly by sodium borohydride to As(III), which is then instantaneously converted to arsine. Determination of total arsenic requires that all inorganic arsenic compounds be in the As(III) state by reduction of any As(V) to As(III) with sodium/potassium iodide, after initial conversion of all inorganic and organic arsenic compounds and standards to As(V) by digestion. Arsine is evolved by reduction with sodium borohydride ($NaBH_4$) from $HNO_3 - H_2SO_4$ soil digest media. Ar carrier gas sweeps the arsine directly into a flame-heated quartz cell mounted in the optical path of a suitably equipped atomic absorption spectrophotometer.

***Note:*** Certain atomic absorption atomizers and hydride reaction cells are available commercially for use with sodium borohydride reagent. Irrespective of the hydride reaction cell-atomizer system selected, it must meet the following quality control considerations:

- It must provide a precise and reproducible standard curve between 0.20 μg As/L and a detection limit between 0.1 and 0.5 μg As/L.
- When carried through the entire procedure, oxidation state couple [As(III) – As(V)] must cause equal instrumental response.
- Sample digestion must yield 80% or greater recovery of added (dimethyl arsenic acid) and 90% or greater recovery of added As(III), As(V).

*Caution* Arsenic and its hydride is toxic, handle with care.

## Apparatus and reagents

- Atomic absorption spectrophotometers; equipped with gas flow meters for argon (or nitrogen) and hydrogen, As electrode less discharge lamps with power supply, background correction at measurement wavelengths, and appropriate strip-chart recorder.
- Atomizer; Cylindrical quartz cell, 10 to 20 cm long electrically heated by external nichrome wire to 800–900°C. or cylindrical quartz cell with internal fuel rich hydrogen-oxygen (air) flame. The sensitivity of quartz cells deteriorates over several months of use. Sensitivity sometimes may be restored by treatment with 40% HF.
- Reaction cell for producing As hydride; A commercially available system is acceptable if it utilizes liquid sodium borohydride reagents.
- Sodium borohydride reagent; Dissolve 8 g sodium borohydride in 200 mL of 0.1 (N) NaOH. Prepare fresh daily;
- 10% KI; Add 10 g of potassium iodide (KI) and dilute to 100 mL in a flask.

## Procedure

### Digestion for total As

- Pulverize dry soil in an agate mortar to pass non-metallic 100 mesh screen.
- Transfer 1.0 g of the dry, pulverized soil into the Kjeldahl flask, add three glass beads, 30 mL of concentrated $HNO_3$ and swirl to mix contents.
- Place flask on micro-Kjeldahl digester in a vented hood.
- Predigest by slow heating for 45 minutes taking care not to permit severe foaming or bumping. It may also be necessary to rotate flask to prevent soil caking.
- Increase temperature to produce steady boiling and continue process until 2–4 mL of $HNO_3$ remains in flask.
- Remove from the digester and allow to cool.
- Add 5 mL concentrated $H_2SO_4$, swirl to mix, return to digester and heat to boiling.
- After fumes cease, remove from the digester and cool.
- Add 25 mL of saturated ammonium oxalate, swirl to mix and return to digester.
- Continue soil digestion until fuming ceases. Soil digest should be light gray to nearly white on completion of digestion. The entire process takes about 3 hours.

### Arsenic extraction using 0.5 (M) sodium bicarbonate solution adjusted to pH 8.5 (Johnston and Barnard; 1979).

- Take 5 g air dried sample in a conical flask and add 100 mL of 0.5 (M) $NaHCO_3$ (pH 8.5) solution.
- Mix thoroughly and shake on a reciprocating shaker for 18 hours.
- Centrifuge for 10 minutes at 2000 r.p.m. Filter through Whatman No. 42 filter paper.
- This solution is used for arsenic estimation.

### Determination of arsenic with sodium borohydride

To 50 mL of the above extract add 5 mL conc. HCl and 5 mL 10% KI and wait for half an hour. This solution is feed to the AAS coupled with hydride generator together with sodium borohydride, concentrated HCl and the purger gas. (Ar or $N_2$) which transports the generated arsine ($AsH_3$) to the quartz cell aligned with the arsenic hollow cathode lamp (Model : GBC 932B).

### Standard curve

Commercially available Arsenic standard solution 1000 mg/L is available (Merck; Germany). From this standard desired concentrations are prepared using double distilled water.

### Calculation

The As concentration ($\mu g\ g^{-1}$) is obtained directly from the standard curve which is calibrated in the instrument keeping in mind the proper dilution factor.

## 3.23 SELENIUM DETERMINATION BY CONVERSION TO THEIR HYDRIDES AND ASPIRATION INTO AN AAS

Selenium (Se) - a group VIB element is an essential micronutrient for humans and animals, but if ingested in relatively large quantities, it also can be toxic. Concern for Se in soils and the environment is related to both deficiency and toxicity of Se in animals. Deficiency problems generally are related to low Se concentrations in feed and forage plants (< 0.05–0.1 mg $kg^{-1}$ Se) (Sharma and Singh, 1983), whereas toxicity problems usually result from bioaccumulation and biomagnification of Se in the food chain. Ure and Berrow (1982) report a Se concentration range of 0.03 to 2.0 mg $kg^{-1}$ and a mean of 0.40 mg $kg^{-1}$ for 1623 soils from throughout the world.

The geochemistry of Se in the environment (e.g., McNeal and Balistrieri, 1989) is directly related to its occurrence in four oxidation states. The most oxidized and most soluble Se species is selenate [Se(VI), $SeO_4^{2-}$]. Under mildly oxidizing conditions, selenite [Se(IV), $SeO_3^{2-}$] predominates. Elemental Se ($Se^o$) is very insoluble and predominates under reducing conditions. Under very reducing conditions, selenide ($Se^{2-}$) predominates. Insolubility of meal selenides (e.g., Fe, Cu) and volatility of $H_2Se$ generally limit the presence of $Se^{2-}$ in aqueous environments. Selenide also substitutes for $S^{2-}$ in organic molecules such as amino acids (e.g., selenocysteine, selenomethionine) and occurs in methylated forms (e.g., dimethylselenide, dimethyldiselenide) in the environment (Doran and Alexander, 1977; Cooke and Bruland, 1987; Masscheleyn *et al.*, 1991a), as discussed below.

Measuring total Se in soils generally is a first approach used to identify problems related to toxicity or deficiency.

### Total selenium in soil

The most common methods used to determine total Se in soil require initial decomposition of the soil followed by analysis of total Se in the resulting solution. Methods to decompose soil samples involve a multitude of mineral acids and various heating techniques to digest the soil. Analysis of Se in solutions resulting from sample digestion and in aqueous samples has been accomplished by many methods. However, the Hydride Generation Atomic Absorption Spectrometric (HGAAS) method is described in detail below.

**Digestion of soils:** The recommended procedure for decomposing soil to determine total Se involves digestion of soil in mineral acids at elevated temperatures. This method involves digesting soil samples with a mixture of nitric acid ($HNO_3$), hydrochloric acid (HCl), perchloric acid ($HClO_4$), sulfuric acid ($H_2SO_4$), and hydrofluoric acid (HF), followed by addition of HCl to reduce Se(VI) to Se(IV). For samples containing relatively high concentrations of organic matter (greater than about 40 g $kg^{-1}$ organic C), samples should stand overnight at room temperature after the initial addition of $HNO_3$ and HCl.

### Method

Digestion of Soil

### Reagents

- Nitric acid, concentrated, 16 M.

- Hydrochloric acid, 1.2 M, prepared by dilution of concentrated HCl, 12 M.
- Hydrochloric acid, 6 M, prepared by dilution of concentrated HCl, 12 M.
- Perchloric acid, concentrated.
- Sulfuric acid, concentrated, 18 M.
- Hydrofluoric acid, concentrated, 48%.

## Procedure

- Weigh a 0.250-g sample (< 0.18 mm) into a 30-mL Teflon bottle and add 9 mL of concentrated $HNO_3$ and 0.25 mL of 1.2 M HCl.
- If the sample contains relatively elevated organic matter (greater than about 40 g $kg^{-1}$), allow to stand overnight at room temperature.
- Add 2 mL of concentrated $HClO_4$, 2 mL of concentrated $H_2SO_4$, 10 mL of concentrated HF, and heat overnight at 125 °C.
- Remove the sample from the heat and let it cool.
- Add 25 mL of 6 M HCl and let the sample stand for 1 h.
- Quantitatively transfer the sample to a 50 mL volumetric flask, and bring to volume with deionized distilled water for HGAAS analysis.

## Selenium determination (Hydride generation atomic absorption spectrometry)

## Reagents

- Hydrochloric acid. 7 M: Prepare by dilution of concentrated HCl, 12 M.
- Sodium borohydride solution, 0.6% (w/v): Prepared fresh daily by dissolving 2.50 g sodium hydroxide (NaOH) in deionized distilled water, adding 3.00 g $NaBH_4$, and diluting to a total volume of 500 mL with deionized distilled water.
- Selenium standard solutions: Prepare from commercially available AAS standard (1000 mg $L^{-1}$ Se) by making serial dilutions to concentrations of 2, 5, 12, and 18 μg $L^{-1}$ Se in 1.2 M HCl. Standard solutions should be made for each set of analyses and should be treated the same as the solutions resulting from the soil digestion procedure. HGAAS method uses sodium borohydride ($NaBH_4$) under acidic conditions to reduce Se(IV) and generate $H_2Se$.

$$3NaBH_4 + 4H_2SeO_3 \longrightarrow 4H_2SeO_3(g) + 3H_3BO_3 + 3NaOH \quad ...(3.23.1)$$

The resulting $H_2Se$ is flushed from solution with an inert gas into a heated quartz absorption cell mounted in the light path of an AAS, where Se absorbance is measured.

The HGAAS method uses a continuous-flow hydride generator coupled to an AAS to measure total Se in the solution resulting from soil digestion. The automated continuous-flow hydride generators are designed to combine the sample, the $NaBH_4$ reagent, and the 7 M HCl reagent to form the $H_2Se$ [Eq.(3.17)]. The mixture goes through a gas/liquid separator, an inert gas, such as Ar, carries the $H_2Se$ into the quartz absorption cell mounted in the light path of the

AAS where Se is ionized, and its absorbance is measured at a wavelength of 196.0 nm. Heating of the quartz absorption cell is commonly accomplished using an air-acetylene flame, but more uniform heating and better precision have been reported for electrically heated quartz absorption cells (Chapman and Dale, 1970).

The absorbance of the sample is compared to the standard curve to determine the total Se concentration in micrograms per liter. This value is multiplied by any dilution factor resulting from dilution of the original digestion solution, and by 0.2 to obtain the concentration of total Se in soil in milligrams per kilogram. This method has a practical detection limit of 0.2 mg $kg^{-1}$ Se.

## Available selenium

Sequential selective-extraction methods recently have been developed specifically for fractionating Se in soils (Chao and Sanzolone, 1989). The method developed by Chao and Sanzolone (1989) determines Se in five operationally defined soil fractions. General information regarding extractant, forms extracted, mechanisms of extraction, and relative availability index for each fraction are listed in Table 3.23.1.

**Table 3.23.1** General information on the partial dissolution techniques

| Extractant | Forms of extracted or dissolved | Mechanisms involved | Availability for leaching and plant uptake† |
|---|---|---|---|
| 0.25 M KCl | $H_2O$ soluble, nonspecifically adsorbed selenate | Mass action, anion exchange | Highly available |
| 0.1 M $KH_2PO_4$ | Exchangeable, specifically adsorbed selenite | Anion competition, ligand exchange | Available |
| 4 M HCl | Associated with oxide minerals (Fe, Mn, and Al and amorphous materials), carbonates, acid-volatile sulfides, and acid hydrolizable organic matter | Acid dissolution, acid hydrolysis | Conditionally available |
| $KClO_{4+}$ concentrated HCl | Sulfides, complex humified organic matter | Oxidative acid dissolution | Unavailable |
| HF + $HNO_3$ + $HClO_4$ | Siliceous materials | Acid destruction of mineral -lattice structure | Resistant |

† Availability for leaching and plant uptake based was on the conceptual relative rating between being highly available and being resistant.

The method described in "Total Selenium" for selenium determination can be used to measure selenium extracted by various extraction methods.

## 3.24 FLUORIDE ESTIMATION IN SOIL AND WATER: SPADNS METHOD

[Sodium 2 – (Parasulfo phenylazo) – 1,8 – dihydroxy – 3,6 – naphthalene disulfonate] also called [4,5 dihydroxy – 3 – (para-sulfophenylazo) – 2,7 – napthalenedisulfonic acid trisodium salt].

Fluorine is common in terrestrial environment and is always present in plants, soils and phosphatic fertilizers. As a rule of thumb, the F concentrations in these materials are on the order of $3\times10^0$, $3\times10^2$ and $3\times10^4$ ppm for plants, soils and phosphatic fertilizers, respectively (Larsen and Widdowson, 1971). Fluorine is a common constituent of rocks and soils. Very common soil minerals, such as biotite, muscovite, and hornblende, may contain as such as several percent of F and, therefore, would seen to be the main source of F in soils. However, Steinkoenig (1919) believed that micas, apatite, and tourmaline in the parent materials were the original source of F in soils. It appears, therefore, that the F content of soils is largely dependent on the mineralogical composition of the soil's inorganic fraction. Phosphatic fertilizers, especially the superphosphates, are perhaps the single most important source of F in agricultural lands.

Fluorine is not an essential plant nutrient but is essential for animals. However, continuous ingestion by the animals of excessive amounts of F can lead to fluorosis, and sub optimal levels in the diet, can have an equally damaging effect. Therefore, plant content of F is of interest to livestock producers. It has been conclusively shown that plant uptake of F from soil is not usually related to the F content of the soil but rather is largely dependent on the soil type, Ca and P contents, and soil pH (Brewer, 1965b).

Fluorine is found in soils as the singly charged fluoride ion, $F^-$ or occasionally as a component of such complex anions as $(BF_4^-)$, $(AlF_6)^{3-}$ and $(S_2F_6)^{2-}$ (Hopkins, 1977).

The problem of high fluoride concentration in groundwater resources has now become one of the most important health-related geo-environmental issues in India, since it has considerable impact, on human physiology. Its deficiency (< 0.6 mg/L) causes dental caries and excess (>1.5 mg/L) causes skeletal fluorosis, dental flurosis respectively. The weathering and leaching processes, mainly by moving and percolating water play an important role in the incidence of fluoride in groundwater. The various factors that govern the release of fluoride into water by the fluoride bearing minerals are (*i*) the chemical composition of water, (*ii*) the presence and accessibility of fluoride minerals to water, and (*iii*) the time of contact between the source mineral and water.

### Principle

This is a colorimetric method and colour development is virtually instantaneous and no waiting is required before measuring fluoride concentration. Colour determinations are made photometrically, using a spectrophotometer. A curve developed from standards can be used for determining the fluoride concentration of a sample or the concentration can be calculated on the basis of a pair of standards. The latter technique makes use of the fact that the relationship between fluoride concentration and absorbance (within the range of the method) is linear and thus that two points can define accurately the position of the line.

### Apparatus and reagents

- Spectrophotometer for use at 570 nm providing a light path of at least 1 cm.
- Standard fluoride solution

(*i*) Stock fluoride solution: Dissolve 221.0 mg anhydrous sodium fluoride, NaF in distilled water and dilute to 1000 mL; 1 mL = 100 μg $F^-$.

(*ii*) Standard fluoride solution: Dilute 100 mL stock fluoride solution to 1000 mL with distilled water.

$$1.00 \text{ mL} = 10.0 \ \mu\text{g } F^-.$$

- SPADNS solution: Dissolve 958 mg SPADNS, [Sodium 2 – (parasul-fophenylazo) – 1,8 – dihydroxy – 3,6 – naphthalene disulfonate] in distilled water and dilute to 500 mL. This solution is stable for at least 1 year if protected from direct sunlight.
- Zirconyl-acid reagent: Dissolve 133 mg zirconyl chloride octahydrate $ZrOCl_2 \cdot 8H_2O$, in about 25 mL distilled water.
- Acid zirconyl – SPADNS reagent; Mix equal volumes of SPADNS solution and zirconyl-acid reagent. The combined reagent is stable for at least 2 years.
- Reference solution: Add 10 mL SPADNS solution to 100 mL distilled water. Dilute 7 mL conc. HCl to 10 mL and add to the diluted SPADNS solution. The resulting solution, used for setting the instrument reference point (zero), is stable for at least one year. Alternately, use a prepared standard of 0 mg $F^-$/L as a reference.
- Sodium arsenite solution: Dissolve 5 g of $NaAsO_2$ and dilute to 1 L with distilled water. Avoid ingestion.

## Procedure

### Preparation of standard curve

Prepare fluoride standards in the range of 0 to 1.40 mg $F^-$/L by diluting appropriate quantities of standard fluoride solution to 50 mL with distilled water. Pipette 5 mL each of SPADNS solution and zirconyl-acid reagent or 10 mL mixed acid-zirconyl SPADNS reagent, to each standard and mix well. Set spectrophotometer to zero absorbance with the reference solution and obtain absorbance readings of the standards. Plot the curve of the mg fluoride-absorbance relationship. Prepare a new standard curve whenever a fresh reagent is prepared. As a alternative to using a reference, set photometer at some convenient point (0.300 or 0.500 absorbance) with the prepared zero mg $F^-$/L standard.

### Colour development

Use a 50 mL sample or a portion diluted to 50 mL with distilled water. Adjust sample temperature to that used for the standard curve. Add 5 mL each of SPADNS solution and zirconyl-acid reagent or 10 mL acid-zirconyl-SPADNS reagent. Mix well and read absorbance first setting the reference point of the photometer as above. If the absorbance falls below the range of standard curve, repeat using a diluted sample.

### Calculations

$$\text{mg } F^-/\text{L} = \frac{A}{\text{mL sample}} \times \frac{B}{C}$$

where A = µg $F^-$ determined from plotted curve

The ratio (B/C) applies only when a sample is diluted to a volume B, and a portion C is taken from it for colour development,when the prepared 0 mg $F^-$/L standard is used to set the photometer.

Alternatively calculate fluoride concentration as follows:

$$\text{mg F}^-/\text{L} = \frac{X - Y}{X - Z}$$

where X = absorbance of the prepared 0 mg $F^-$/L standard,

Y = absorbance of the prepared sample,

Z = absorbance of the prepared 1.0 mg $F^-$/L standard.

## 3.25 DETERMINATION OF LIME REQUIREMENT OF SOIL (SCHOEMAKER *et al.* 1961)

For satisfactory plant growth, the soil should have a pH between 6.5 and 7.5, though certain plants can grow well at low pH like tea and also at high pH like sugarbeet. In India acid soils are located mostly in eastern, southern and south central parts. Also some soils at higher elevations in north India are acidic. For sustained agricultural production and higher yields, through efficient soil management practices, it is essential to lime and acid soil, as it has considerable influence on soil environment, besides correcting soil acidity.

### Principle

In this method the soil is equilibrated with a pH 7.5 buffer solution, whereby the reserve $H^+$ is brought into solution, which results in the depression of pH of the buffer solution, a note of which is made and interpreted in terms of lime required to raise the pH to a desired value.

### Reagents

- Extractant buffer: Dissolve 1.8 g paranitrophenol, 3 g potassium chromate, 2 g calcium acetate, 53.1 g calcium chloride dihydrate ($CaCl_2 \cdot 2H_2O$) and 2.5 mL triethanolamine in 1 litre of distilled water. Adjust the pH to 7.5 with NaOH.

### Procedure

- Determine the pH of the soil sample in 1 : 2.5 soil water ratio.
- For this weigh 10 g soil and add 25 mL distilled water.
- Shake intermittently for half an hour and record the soil pH. If the pH exceeds 6.0 then this method is not applicable.
- If the measured pH is 6.0 or low then proceed as follows:
- Weigh 5 g soil in a 50 mL beaker.
- Add to it 5 mL of distilled water and 10 mL buffer solution.
- Stir continuously for 10 minutes or intermittently for 20 minutes.

- Determine the soil pH with the pH meter.
- Lime requirement is determined on the basis of soil-buffer pH ready reckoner given below.

The values in this table are given in tons of pure $CaCO_3$ per acre required to bring the soil to the indicated pH and thus are required to be converted to their equivalents in the form of agricultural lime to be used. The figures are multiplied by a factor of 2.43 to express in tons per hectare.

| pH of soil-buffer suspension | Lime required to bring the soil to indicated pH (tons/acre of pure $CaCO_3$) | | |
|---|---|---|---|
| | pH 6.0 | pH 6.4 | pH 6.8 |
| 6.7 | 1.0 | 1.2 | 1.4 |
| 6.6 | 1.4 | 1.7 | 1.9 |
| 6.5 | 1.8 | 2.2 | 2.5 |
| 6.4 | 2.3 | 2.7 | 3.1 |
| 6.3 | 2.7 | 3.2 | 3.7 |
| 6.2 | 3.1 | 3.7 | 4.2 |
| 6.1 | 3.5 | 4.2 | 4.8 |
| 6.0 | 3.9 | 4.7 | 5.4 |
| 5.9 | 4.4 | 5.2 | 6.0 |
| 5.8 | 4.8 | 5.7 | 6.5 |
| 5.7 | 5.2 | 6.2 | 7.0 |
| 5.6 | 5.6 | 6.7 | 7.7 |
| 5.5 | 6.0 | 7.2 | 8.3 |
| 5.4 | 6.5 | 7.7 | 8.9 |
| 5.3 | 6.9 | 8.2 | 9.4 |
| 5.2 | 7.4 | 8.6 | 10.0 |
| 5.1 | 7.8 | 9.1 | 10.6 |
| 5.0 | 8.2 | 10.1 | 11.2 |
| 4.9 | 8.6 | 10.6 | 11.8 |
| 4.8 | 9.1 | | 12.4 |

## 3.26 DETERMINATION OF GYPSUM REQUIREMENT OF SOIL

### Principle

Presence of large amount of sodium as high as 15% or more in the exchange complex results in high soil pH ( > 8.0) for sodic (alkali) and saline-sodic soils which causes nutritional imbalances, depletion of soil organic matter, deterioration of soil physical health and also affects the soil biotic community Gypsum ($CaSO_4 \cdot 2H_2O$) is commonly used for soil amendment under such situation. A fixed weight of soil is equilibrated with a known amount of Ca solution, and the amount of Ca left in solution is determined by EDTA-titration. The difference between the amount of Ca added and Ca left in solution, gives the amount of Ca exchanged. Practically it has been observed that gypsum addition of about 1/3 of the value obtained by this method is satisfactory in most cases.

## Reagents

- Ammonium chloride – Ammonium hydroxide buffer: Dissolve 67.5 g of $NH_4Cl$ in 570 mL of $NH_4OH$ (sp.gr.0.86), and dilute to 1 litre.
- Saturated $CaSO_4$ solution: Shake about 5 g $CaSO_4 \cdot 2H_2O$ with 1 litre of distilled water for 15 minutes on a mechanical shaker and filter.
- Eriochrome Black T (EBT) indicator: Dissolve 0.5 g of EBT and 4.5 g of hydroxylamine hydrochloride in 100 mL of 95% ethanol.
- Standard EDTA solution 0.01N: Dissolve 2 g of disodium dihydrogen-ethylene-diamine-tetra acetate and 0.05 g of $MgCl_2 \cdot 6H_2O$ in water and dilute to 1 litre. Standardise against standard Ca-solution.

## Procedure

- Weigh 5 g of soil sample in a 250 mL conical flask, and add 100 mL of saturated $CaSO_4$ solution.
- Shake for 5 minutes on a mechanical shaker and filter.
- Pipette out 5 mL of the extract into a 100 mL conical flask and dilute to about 25 mL with distilled water.
- Add 0.5 mL of the $NH_4Cl$-$NH_4OH$ buffer and 3–4 drops of the EBT indicator.
- Titrate with the standard EDTA solution until the colour changes from wine red to blue.
- Titrate in a similar way 5 mL of the saturated $CaSO_4$ separately.

## Calculations

$$\text{Weight of soil taken} = 5\text{ g}$$

$$\text{Total volume of extract} = 100\text{ mL}$$

Let volume of EDTA used for titration of $x$ mL of gypsum solution be A mL (say) and volume of EDTA used for titration of $y$ mL of sample aliquot be B mL (say)

Therefore, meq. of Ca/L in gypsum solution

$$= \left(\frac{A}{X}\right) \times 0.01 \times 1000 = P \text{ meq./L}$$

Meq. of Ca/L in sample solution

$$= \left(\frac{B}{Y}\right) \times 0.01 \times 1000 = Q \text{ meq./L}$$

Total meq. of Ca remained in soil after addition of 100 mL gypsum

$$= \frac{(P-Q)\times 100}{1000} = 0.1\ (P - Q)$$

Now 5 g soil contains 0.1 × (P – Q) meq.

Hence 100 g soil contains $\frac{0.1\times(P-Q)}{5} \times 100$ meq.

$$= [2\times(P - Q)]\ \text{meq./100 g}$$

Thus 1 kg soil requires $= [20\times 20\times(P - Q)]$ mg Ca

$$= 400 \times (P - Q)\ \text{mg Ca}$$

2.24 million kg soil requires $= \frac{400\times(P-Q)\times 2.24\times 10^6}{10^6}$

$$= 896\ (P - Q)\ \text{kg Ca.}$$

Now 40 kg Ca is obtained from 172 kg gypsum

So, [896 × (P – Q)] kg Ca is obtained from $\frac{172\times 896}{40\times 1000} \times (P - Q)$

$$= [3.85\times(P - Q)]\ \text{tons of gypsum}$$

Thus gypsum requirement of the soil = [3.85 × (P – Q)] tons/ha

## 3.27 DETERMINATION OF LIME POTENTIAL

Soil pH measurement is usually affected by a number of factors viz. salt concentration, soil-water ratios, suspension effect etc. The use of 0.01(M) $CaCl_2$ solution yields stable readings in pH-measurements. The hydrogen ions in the soil system are distributed between solid and liquid phases as follows. The dynamic equilibrium may be represented as follows.

$$H^+_{\text{Exch}} \rightleftharpoons H^+_{\text{Soln.}} \quad \text{...(3.27.1)}$$

The $H^+$ ions in the soil solution constitutes the 'active acidity' and are measured directly as soil pH values. On the other hand the adsorbed $H^+$ ions held on exchange sites are not subject to pH measurements are termed as **reserve acidity**, both the forms contribute to soil acidity. Thus soil pH does not reflect the total acidity. However a suitable index which takes into account the reserve acidity of soil is the lime potential, which is calculated as follows:

$$\text{Lime potential} = \text{pH} - \tfrac{1}{2}\,\text{pCa} \quad \text{...(3.27.2)}$$

where, $$\text{pCa} = -\log\ (Ca^{2+}) \quad \text{...(3.27.3)}$$

The lime potential is a very reliable estimate to predict the buffering capacity of soils. Smaller the value of lime potential, greater will be the buffering capacity of the soil.

Schofield and Taylor (1955) suggested the use of ion acitivity ratios for determination of soil acidity. Consider a cation C with valency ν. In dilute solution in equilibrium with a soil the acitivity of hydrogen ions divided by the activity of C ions will be constant. The activity function depends on the valency of ions concerned.

Therefore $$\frac{a_{H+}}{(a_C)^{1/v}} = \text{constant} \qquad ...(3.27.4)$$

If the soil exchange complex is saturated with both $H^+$ and $Ca^{2+}$ ions at equilibrium. Schoefield's ratio law states:

$$\frac{a_{H^+}}{\sqrt{a_{Ca}^{2+}}} = \text{constant (K)} \qquad ...(3.27.5)$$

The negative logarithm of (K) is called the lime potential.

$$i.e., \ -\log \frac{a_{H^+}}{\sqrt{a_{Ca}^{2+}}} = -[\log a_{H^+} - \log \sqrt{a_{Ca}^{2+}}\,] = -[\log a_{H^+} - \frac{1}{2}\log a_{Ca}^{2+}] \qquad ...(3.27.6)$$

$$= (-\log a_{H^+} + \frac{1}{2}\log a_{Ca}^{2+}) = [(-\log a_{H^+}) - \frac{1}{2}(-\log a_{Ca}^{2+})] = (\text{pH}-\frac{1}{2}\text{pCa}) \qquad ...(3.27.7)$$

### Principle

In practice soil is shaken with a calcium chloride solution of known strength (1 : 2 soil solution ratio) and activity of $Ca^{2+}$ ions and the pH of the suspension is measured. Lime potential is calculated as; (measured pH – 1.14), 1.14 being the value of ½ pCa for 0.01(M) $CaCl_2$ solution. The use of 0.01(M) $CaCl_2$ solution as an extractant simulates the electrolye level of non-saline soil at optimum field water content and more so the $H^+$ ion environment existing in the soil solution-plant root system.

### Reagents

- 0.01(M) $CaCl_2$ solution: Dissolve 1.3 g of anhydrous $CaCl_2$ in water and dilute to one litre.
- Buffer solutions for pH measurement.

### Procedure

- Weigh accurately 10 g of soil in a conical flask and add 20 mL $CaCl_2$ extracting solution into it.
- Shake for half an hour and measure the pH of the suspension using a pH meter.

### Calculation

Lime potential = pH – 1.14

## 3.28 AVAILABLE SULPHUR DETERMINATION IN SOIL

Turbiditmetric Procedure; Monocalcium Phosphate Extratable S (Ensminger, 1954)

### Principle

Sulphur (S) occurs in soils usually as sulphites, sulphates, sulphides and in organic compounds. However, the most accessible form is 'sulphate' ($SO_4^{2-}$). The turbidimetric procedure is widely used in the estimation of available S in the soil due to its rapidity. However, erroneous results are obtained in case the soil is rich in organic matter. Soil is shaken with a solution of monocalcium phosphate, containing 500 ppm P. The phosphate ions displaces the adsorbed sulphate. The calcium ions depresses the extraction of soil organic matter, thus eliminating contamination from extractable organic S. The method extract soluble $SO_4^{2-}$ plus a fraction of adsorbed $SO_4^{2-}$. The filtrate is then analysed for S by the turbidimetric procedure. In this method the filtrate is treated with barium chloride in the presence of gumacacia solution, and the turbidity produced by the precipitation of sulphate as barium sulphate is measured colorimetrically. Gumacacia helps in preventing rapid settling of barium sulphate precipitate.

$$Ca(H_2PO_4)_2 \cdot 2H_2O \longrightarrow \overset{+}{Ca} + 2H_2PO_4^- + 2H_2O \quad ...(3.28.1)$$

$$H_2PO_4$$

$$\text{Soil colloid] } SO_4 + 2H_2PO_4^- \longrightarrow \text{Soil colloid ] } + SO_4^{2-} \quad ...(3.28.2)$$

$$H_2PO_4$$

$$BaCl_2 \longrightarrow Ba^{2+} + 2Cl^- \quad ...(3.28.3)$$

$$Ba^{2+} + SO_4^{2-} \longrightarrow \underset{\text{precipitate}}{Ba(SO_4)} \quad ...(3.28.4)$$

## Reagents

- Monocalcium phosphate solution; Dissolve 2.18 g of $Ca(H_2PO_4)_2 \cdot 2H_2O$ in distilled water, and dilute to one litre.
- Barium chloride: Grind $BaCl_2$ crystals in a mortar, until they pass through a 30 mesh sieve and are retained on a 60 mesh sieve. The crystals are added to sulphate solution in the solid state as crystals of definite size and not as solution.
- Gum-acacia solution, 0.25%: Dissolve 0.25 g gum-acacia in distilled water, and dilute to 100 mL.
- Standard S solution: Dissolve 0.5434 g (AR-grade) potassium sulphate in distilled water and dilute to one litre. This gives 100 ppm stock solution of S.

## Procedure

- Weigh accurately 20 g of air dried soil and transfer it to a 250 mL Erlenmeyer flask.
- Add 100 mL of monocalcium phosphate extracting solution and shake for half an hour. Filter through Whatman No. 42 filter paper under suction.
- Take 20 mL of the filtrate quantitatively in a 25 mL volumetric flask and then proceed as described in the preparation of standard curve.

## Standard curve

- Pipette out 0, 2.5, 0.5, 1.0, 2.0, 2.5, 5.0 mL of the 100 ppm stock solution in a series of 25 mL volumetric flask.
- Add to each flask 10 mL of the monocalcium phosphate solution followed by 1g $BaCl_2$ crystals and shake for 1 minute.

- Add 1 mL of 0.25% gum-acacia. Make up the volume of each flask with distilled water and shake thoroughly for 1 minute. These are working S standards (1, 2, 4, 8, 10, 20 ppm S, respectively).
- Make turbidity measurements, following formation of precipitate from 5–30 minutes with the help of a spectrophotometer at 420 mμ wavelength.
- Plot a curve showing turbidity readings (absorbance as ordinate and concentration of S in ppm as abscissa).

### Calculations

$$\text{Weight of soil taken} = 20 \text{ g}$$

$$\text{Volume of extractant added} = 100 \text{ mL}$$

$$\text{Hence, first dilution} = \frac{100}{20} = 5 \text{ times}$$

$$\text{Volume of aliquot taken} = 20 \text{ mL}$$

$$\text{Final volume} = 25 \text{ mL}$$

$$\text{Hence, second dilution} = \frac{25}{20} = 1.25 \text{ times}$$

$$\text{Thus total dilution} = 5 \times 1.25 = 6.25 \text{ times}$$

Let ppm of sulphur obtained from standard curve be S (say)

Now, available S in soil (ppm) = $S_1 \times 6.25$

or Available S = $(S_1 \times 6.25 \times 2.24)$ kg/ha.

## 3.29 DETERMINATION OF CARBONATE ($CO_3^{2-}$) AND BICARBONATE ($HCO_3^-$) IN SOIL

### Principle

When phenolphthalein is used as an indicator, strong alkalis like KOH or NaOH are completely neutralized whereas weak alkalis like $Na_2CO_3$ or $K_2CO_3$ are neutralized to the stage of $NaHCO_3$ or $KHCO_3$ according to the equation.

$$Na_2CO_3 + H_2SO_4 \longrightarrow NaHSO_4 + NaHCO_3 \quad ...(3.29.1)$$

The $NaHCO_3$ thus formed requires more $H_2SO_4$ to get completely neutralized according to the equation.

$$2NaHCO_3 + H_2SO_4 \longrightarrow Na_2SO_4 + 2CO_2 + 2H_2O \quad ...(3.29.2)$$

It is evident from the above equations that the quantity of $H_2SO_4$ required in both the stages of neutralization of $Na_2CO_3$ is the same. The second stage of neutralization of $Na_2CO_3$ (*i.e.*, the neutralization of $NaHCO_3$) can be indicated by methyl orange which can also indicate complete neutralization of alkali carbonate or bicarbonate. Thus phenolphthalein and methyl red are used one after the other during the course of titration in the same solution for evaluating mixtures containing carbonates and bicarbonates. Methyl orange when used jointly with phenolphthalein after the latter has decolourised indicates the quantity of acid required for the neutralization of the bicarbonate only.

## Reagents

- Phenolpthalein indicator; 0.25% solution in 60% ethylalcohol.
- Methyl orange indicator; 0.5% solution in 95% alcohol.
- Standard $H_2SO_4$; 0.01 (N)

## Procedure

- Weigh 40 g of soil sample in a 500 mL conical flask.
- Add 200 mL double distilled water and shake for one hour in a shaking machine for equilibration.
- Filter the suspension.
- Pipette out 5 mL of the extract or 5 mL of water sample (containing not more than 1 meq. of $CO_3^{2-}$ plus $HCO_3^-$) in a porcelain dish and add 2–3 drops of phenolphthalein indicator. Titrate against 0.01(N)$H_2SO_4$ until the pink colour just disappears (indicating phenolphthalein end point). This end point corresponds to the neutralization of the carbonate to the bicarbonate stage.
- Record the mL of 0.01(N) $H_2SO_4$ required for this process from the burette reading.
- Add 1–2 drops of methyl orange indicator to the colourless solution.
- Titrate it again with 0.01(N) $H_2SO_4$ stirring briskly, until the indicator turns orange indicating complete neutralization of the bicarbonate present.
- Note the titre value from the burette.

## Calculations

Weight of soil taken = 40 g

Volume of water added = 200 mL

Let volume of aliquot taken from soil extract or water sample be V mL.

Volume of 0.1 (N) $H_2SO_4$ required for the first titration (with phenolphthalein) = $t_1$ mL.

Total volume of $H_2SO_4$ required = $t_2$ mL

(phenolphthalein plus methyl red)

Normality of $H_2SO_4$ used = 0.01 (N) or $N_1$(say)

Therefore meq. of $H_2SO_4$ used in the first titration = $N_1 \times t_1$

meq. of $H_2SO_4$ used (total) in the successive titration = $N_1 \times t_2$

$$\text{Hence meq. of } CO_3^{2-} \text{ per 100 g of soil} = (N_1 \times t_1) \times \frac{200}{V} \times \frac{100}{40}$$

$$\text{and mg of } CO_3^{2-} \text{ per 100 g soil} = (N_1 \times t_1) \times \frac{200}{V} \times \frac{100}{40} \times 30$$

$$\text{Likewise, meq of } HCO_3^- \text{ per 100 g soil} = \{(t_2 - t_1) \times N_1\} \times \frac{200}{V} \times \frac{100}{40}$$

$$\text{and, mg of } HCO_3^- \text{ per 100 g soil} = \{(t_2 - t_1) \times N_1\} \times \frac{200}{V} \times \frac{100}{40} \times 61$$

*Note:* [1 mL of 0.01 (N) $H_2SO_4$ = (0.01 meq. $H_2SO_4$) = 0.00030 g $CO_3^{2-}$ = 0.00061 g $HCO_3^-$]

$$\text{Also meq. of } CO_3^{2-} \text{ per litre of soil extract or water sample} = (N_1 \times t_1) \times \frac{1000}{V}$$

$$\text{and, meq. of } HCO_3^- \text{ per litre of soil extract or water sample} = \{(t_2 - t_1) \times N_1\} \times \frac{1000}{V}$$

## 3.30 DETERMINATION OF CHLORIDE ($Cl^-$) IN SOIL EXTRACT

### Principle

Chloride determination is based on the formation of nearly insoluble silver salts.

$$Cl^- + Ag^+ \longrightarrow AgCl \quad \text{(white spongy precipitate).} \qquad ...(3.30.1)$$

Silver nitrate in presence of potassium chromate indicator is used for precipitating $Cl^-$,

$$NaCl + AgNO_3 \longrightarrow AgCl + NaNO_3 \qquad ...(3.30.2)$$

$$K_2CrO_4 + 2AgNO_3 \longrightarrow Ag_2CrO_4 + 2KNO_3 \quad \text{reddish brown ppt.} \qquad ...(3.30.3)$$

### Reagents

- Potassium Chromate indicator; 5% aqueous solution of pure $K_2CrO_4$
- 0.02(N) $AgNO_3$ solution; Dissolve 3.4 g of $AgNO_3$(A.R) is double distilled water and make up the volume to one litre. Standardize this solution against a standard NaCl solution and keep in amber coloured bottle away from light.

### Procedure

- Pipette out 50 mL aliquot from the same soil-water extract as that used in $CO_3^{2-}$ and $HCO_3^-$ estimation or 5 mL of the filtered water sample.
- Add 5–6 drops of $K_2CrO_4$ indicator and titrate the solution with 0.02 (N) $AgNO_3$ solution with stirring until the first reddish brown tinge appears. The mL of $AgNO_3$ required corresponds to the amount of chloride present.

### Calculations

Let volume of aliquot taken (from soil extract or water sample) = V mL

Volume of $AgNO_3$ solution used in titration = T mL

Normality of $AgNO_3$ = 0.02 (N) or $N_A$

Therefore, meq. of $AgNO_3$ used in titration = $N_A \times T$

$$\text{meq. of Cl}^- \text{ per litre of soil extract or water sample} = (N_A \times T) \times \frac{1000}{V}$$

$$\text{meq. of of Cl}^- \text{ per 100 g soil} = (N_A \times T) \times \frac{200}{V} \times \frac{100}{40}$$

$$\text{and mg of Cl}^- \text{ per 100 g soil} = (N_A \times T) \times \frac{200}{V} \times \frac{100}{40} \times 35.5$$

***Note:*** 1 mL of 0.02 N $AgNO_3$ = (0.02 meq. $AgNO_3$) = 0.00071 g Cl

●●●

# 4 Fundamental Concepts of Analytical Chemistry

## 4.1 EQUILIBRIUM: THE LAW OF MASS ACTION

The concept of equilibrium is really dynamic and not static, in the sense that when equilibrium is attained the reaction proceeds both in forward and backward directions at equal rates, so that the amount of reactants disappearing per unit time is reproduced from the action in opposite direction. The reactions proceeding in both directions are called 'reversible' reactions which is indicated with double arrows in opposite direction. It is very likely that all chemical reactions are reversible, but in some cases the extent of backward reaction is so small as to be negligible and such reactions are said to proceed to completion in one direction. Under such condition the equilibrium is attained at an extreme end of the concentrations of the resultants, the concentration of unreacted materials being extremely small to be detected.

A quantitative relation between the amounts of reactants and the resultants of equilibrium was developed from a basic principle, called the law of Mass Action enuntiated by two Norweigian chemists, Gulberg and Waage (1867).

**The law states:** Temperature remaining constant, the rate of a chemical reactions is proportional to the 'active masses' of the reacting substances. The expression 'active mass' in the statement requires serious consideration. For solids and pure liquids, the active masses or concentrations are taken as unity since the rate of reaction is independent of their amounts present. Let us consider the simple reversible reaction at constant temperature:

$$A + B \underset{k_2}{\overset{k_1}{\rightleftharpoons}} C + D$$

Let C′–terms denote the concentrations of the components at a given instant during the progress of reaction. According to the law, the rate of the forward reaction ($R_{AB}$) between A and B at that moment will be

$$R_{AB} = k_1 \, C'_A \, C'_B \qquad ...(4.1.1)$$

where $k_1$ is the proportionality constant, depends upon reactants and temperature. The concentrations of the reactants (A and B) would diminish with progress of the reaction and hence its rate would diminish with time. Similarly, the rate of opposite (backward) reaction ($R_{CD}$) between C and D at that moment will be

$$R_{CD} = k_2 \, C_C' \cdot C_D' \qquad ...(4.1.2)$$

With progress of reaction, the concentrations of C and D would increase and hence the rate $R_{CD}$ would increase with time. A time will come when the rates of reaction in the two opposite directions would be equal, *i.e.*,

$$R_{AB} = R_{CD} \qquad ...(4.1.3)$$

The system would then attain equilibrium and there would be no further change in the masses of the components of the system. At equilibrium let the concentrations be expressed by C-terms (instead of C′). Then

$$k_1\ C_A\ C_B = k_2\ C_C\ C_D \text{ (since } R_{AB} = R_{CD}) \qquad ...(4.1.4)$$

or

$$\frac{C_C \cdot C_D}{C_A \cdot C_B} = \frac{k_1}{k_2} = K_C \text{ (Constant)} \qquad ...(4.1.5)$$

$K_c$ is called the equilibrium constant of the reaction.

The expression may be generalised for a reversible reaction represented by;

$$p_1A_1 + p_2A_2 + p_3A_3 + \cdots = q_1B_1 + q_2B_2 + q_3B_3 + \cdots$$

$$k = \frac{(C_{B1})^{q1} \times (C_{B2})^{q2} \times (C_{B3})^{q3} \times \cdots}{(C_{A1})^{p1} \times (C_{A2})^{p2} \times (C_{A3})^{p3} \times \cdots} \qquad ...(4.1.6)$$

Thus it may be stated that the ratio of the product of molecular concentration of the resultant to the product of molecular concentration of the reactance, each concentration being raised to the proper power equal to the number of molecules taking part in the reaction at constant temperature under equilibrium is a constant, called the **equilibrium constant**.

## 4.2 ACTIVITY AND ACTIVITY COEFFICIENT

In the deduction of law of mass action it was assumed that the effective concentrations or the active masses of the components may be expressed by the stoichiometric concentrations. According to modern thermodynamics, this is not strictly true. The rigorous equilibrium equation for a reaction of the type AB $\rightarrow$ $A^+ + B^-$ is given by

$$K_a = \frac{(a_{A^+}) \times (a_{B^-})}{a_{AB}} \qquad ...(4.2.1)$$

where $a_{A^+}$, $a_{B^-}$ and $a_{AB}$ represent the activities of $A^+$, $B^-$ and AB respectively and $K_a$ is the true or thermodynamic dissociation constant.

$$\text{activity} = \text{activity coefficient} \times \text{concentration} \qquad ...(4.2.2)$$

Also for dilute solutions activity is equal to concentration since activity coefficient is generally taken to be unity.

Thus at any concentration,

$$a_{A^+} = \gamma_{A^+} \cdot C_{A^+},\ a_{B^-} = \gamma_{B^-} \cdot C_{B^-} \text{ and } a_{AB} = \gamma_{AB} \cdot C_{AB}.$$

where $\gamma$ refers to the activity coefficients and C-terms denote the concentrations. On substitution into the expression for $K_a$ we get

$$K_a = \frac{(\gamma_{A^+} C_{A^+}) \times (\gamma_{B^-} C_{B^-})}{\gamma_{AB} \cdot C_{AB}} = \frac{C_{A^+} C_{B^-}}{C_{AB}} \cdot \frac{\gamma_{A^+} \cdot \gamma_{B^-}}{\gamma_{AB}} \qquad ...(4.2.3)$$

The activity coefficients of unionised molecules do not differ considerably from unity and for weak electrolytes in which the ionic strength is small, the true or thermodynamic expression reduces to

$$C_{A^+} \cdot C_{B^-} / C_{AB} = K \qquad ...(4.2.4)$$

Ionic strength $I = ½\, C_i\, Z_i^2$ where $C_i$ = ionic concentration in molalities and $Z_i$ is the valency. Up to I = 0.01 may be considered to be very dilute solutions.

## 4.3 ACID–BASE EQUILIBRIA IN WATER, OSTWALDS DILUTION LAW

Consider the dissociation of a weak electrolyte, such as acetic acid, in dilute aqueous solution:

$$CH_3COOH + H_2O \rightarrow H_3O^+ + CH_3COO^-$$

For simplicity above equation may be written in conventional manner as

$$CH_3COOH \rightarrow H^+ + CH_3COO^-$$

Applying law of mass action:

$$K = \frac{C_{CH_3COO^-}\, C_{H^+}}{C_{CH_3COOH}} \qquad ...(4.3.1)$$

where K is the equilibrium constant at a particular temperature and is usually known as the ionisation or dissociation constant.

Hence the ion produced on dissociation are in equilibrium with the undissociated molecules of weak electrolytes in solution. Thus a weak acid solution of HA will have the equilibrium:

$$\begin{array}{cccc} HA & \rightleftharpoons & H^+ & + & A^- \\ C(1-\alpha) & & \alpha C & & \alpha C \end{array}$$

where C is the concentration and $\alpha$ is the degree of dissociation. Applying law of mass action the dissociation constant of the acid is given by

$$K_a = \frac{C_{H^+}\, C_{A^-}}{C_{HA}} = \frac{\alpha C \cdot \alpha C}{(1-\alpha)C} = \frac{\alpha^2 C}{(1-\alpha)} \qquad ....(4.3.2)$$

This is known as Ostwald's dilution law.

## 4.4 SOLUBILITY PRODUCT

For sparingly soluble salts it is an experimentally observed fact that the product of total molecular concentrations of the ions, coefficient raised as their respective power is a constant at constant temperature. The product $K_s$ is called the solubility product. For a binary electrolyte

$$AB \rightleftharpoons A^+ + B^-$$

$$K_{S(AB)} = C_{A^+} \cdot C_{B^-}$$

In general,

$$ApBq \rightleftharpoons p^{Aq+} + q^{Bp-}$$

$$K_{S(ApBq)} = [C_A{}^{q+}]^p \times [C_B{}^{p-}]^q \qquad \text{...(4.4.1)}$$

When excess of sparingly soluble electrolyte, say silver chloride, is shaken up with water, some of it passes into solution to form a saturated solution of the salt and the reaction appears to cease. Actually the following equilibrium is established:

$$AgCl(solid) \longrightarrow Ag^+ + Cl^-$$

The velocity of the forward reaction, at a particular temperature,

$$v_1 = k_1 \qquad \text{...(4.4.2)}$$

where $k_1$ is a constant and the velocity of the reverse reaction is proportional to the activity of each of the reactants, Hence

$$v_2 = k_2 \; a_{Ag^+} \cdot a_{Cl^-} \qquad \text{.....(4.4.3)}$$

where $k_2$ is another constant.

At equilibrium $\quad v_1 = v_2 \qquad$ ...(4.4.4)

Therefore, $\quad k_1 = k_2 \cdot a_{Ag^+} \cdot a_{Cl^-} \qquad$ ...(4.4.5)

or $\quad k_2 = K_S(AgCl) = k_1 = a_{Ag^+} \cdot a_{Cl^-} \qquad$ ...(4.4.6)

In very dilute solutions the activities may be substituted by concentrations as usual.

## 4.5 STABILITY OF COMPLEXES

The thermodynamic stability of a species is a measure of the extent to which the species will be formed from other species under certain conditions, provided that the system is allowed to reach equilibrium. Consider a metal ion M in solution together with a monodented ligand L, the system may be described by the following stepwise equilibria, in which for convenience the coordinated water molecules are not shown.

$$M + L \rightleftharpoons ML; \qquad K_1 = [ML]/[M][L] \qquad \text{...(4.5.1)}$$

$$ML + L \rightleftharpoons ML_2; \qquad K_2 = [ML_2]/[ML][L] \qquad \text{...(4.5.2)}$$

In this way, for $n$th step;

$$ML(n-1) + L \rightleftharpoons MLn; \qquad Kn = [MLn]/[ML(n-1)][L] \qquad \text{...(4.5.3)}$$

The square brackets denote the concentrations. The equilibrium constants $K_1$, $K_2$,...., $K_m$ are referred to as stepwise equilibrium constants.

Alternatively the equilibria can be expressed as follows:

$$M + L \rightleftharpoons ML; \qquad \beta_1 = [ML]/[M][L]$$

$$M + 2L \rightleftharpoons ML_2; \qquad \beta_2 = [ML_2]/[M][L]^2$$

$$M + nL \rightleftharpoons MLn; \qquad \beta n = [MLn]/[M][L]^n$$

The equilibrium constants $\beta_1$, $\beta_2$, ..., $\beta n$ are called the overall stability constants such that $\beta n = K_1 \times K_2 \times \cdots \times Kn$.

A knowledge of stability constant values is of considerable importance in analytical chemistry, since they provide information about concentrations of various complexes formed by metal in specified equilibrium mixtures — a concept which is invaluable in the study of complexometry.

## 4.6 TITRIMETRY

For use in titrimetric analysis a reaction must satisfy the following conditions:

- There must be a simple reaction that can be expressed by a chemical equation, the substance to be determined must react completely with the reagent in stoichiometric or equivalent proportions.
- The reaction kinetics must be rapid. In certain cases addition of a catalyst increases the speed of reaction.
- There must be a marked change in free energy leading to alteration in some physical or chemical properties of the solution at the end point.
- An indicator must be available which should sharply define the end point of reaction by a change in colour or formation of precipitate etc.

### 4.6.1 Titration

This is the process of determining the volume of a substance (usually of a primary standard solution or a standardised secondary standard solution) required to just complete the reaction with a known volume of another substance. The reagent of known concentration is known as titrant and the substance being titrated is known as titrand.

### 4.6.2 Types of Reactions in Titrimetry

The reactions employed in titrimetric analysis may be broadly divided into two main classes.

- Those in which no change in oxidation states occurs, these are dependant upon combination of ions.
- Oxidation-reduction (redox) reactions involving electron transfer or change of oxidation states.

For simplicity and convenience the above two broad categories is subdivided into four main classes:

### Neutralisation reactions or acidimetry and alkalimetry

The two terms are complementary. They involve determination of strength of acid or an alkali solution by titration against a standard solution of alkali or an acid as the case may be. If the strengths in normality of the alkali and the acid solutions are $S_A$ and $S_B$ respectively and $V_A$ mL of the alkali exactly neutralises $V_B$ mL of the acid then:

$$V_A \cdot S_A = V_B \cdot S_B$$

This is the fundamental equation of acidimetry-alkalimetry. If the strength of one is known, the other can be calculated out.

### Complex formation reaction

These depend upon the combination of ions other than hydrogen or hydroxide ions, to form a soluble slightly dissociated ion or compound. Ethylene diamine tetra acetic acid, mostly as the disodium salt EDTA, is a very important reagent for complex formation. The use of metal ion-indicators has highly enhanced its importance in titrimetry. The subject is discussed in section of calcium and magnesium estimation by EDTA.

### Precipitation reaction

Such reactions depend upon the combination of ions resulting in formation of a simple precipitate as in the titration of silver ion with a solution of a chloride. No change in oxidation state occurs.

### Oxidation-reduction reaction

All reactions involving change in oxidation number or transfer of electrons among the reacting substances fall under this category. The standard solutions are either oxidising or reducing agents. The principal oxidising agents are potassium dichromate, potassium permanganate, potassium iodate, etc. while common reducing agents are sodium thiosulphate, iron (*i*) and tin (II) compounds, chromium (*ii*) chloride or sulphate etc.

### 4.6.3 Strength

Strength of a solution means grams of solute dissolved per litre of solution. Usually it is expressed in terms of normality and molarity.

### 4.6.4 Percentage Strength

Percentage strength means the grams of solute per 100 mL of solution.

### 4.6.5 Standard Solution

The solution of accurately known strength (or concentration) is called standard solution. It contains a known weight of the reagent in a definite volume of solution, concentrations being expressed usually in normality.

The standard solutions are of two types: primary standard and secondary standard.

Primary standard solutions are those solutions which are prepared by accurately weighing a chemically pure substance in a chemical balance and then dissolving in a known volume. The substance should not change its composition either during weighing or in solution during its preparation. The following compounds are generally used for primary standard solutions:

For alkali solution : $Na_2CO_3$(anhydrous) and $Na_2B_4O_7$, $10H_2O$

For acid solution : oxalic acid, succinic acid

For redox system : potassium dichromate, sodium oxalate etc.

Secondary standard solutions are those which cannot be prepared by direct weighing in a chemical balance. These are usually prepared by standardising against some primary standard solution. Examples are solution of NaOH, KOH, HCl, $H_2SO_4$, $KMnO_4$.

The primary standard must be easy to obtain, to purify, to dry (preferably at 110–120°C) and to preserve in a pure state. The substance should be unaltered in air during weighing *i.e.,* it must not be hygroscopic, or oxidised by air, nor affected by carbon dioxide. The composition of primary standard must remain unchanged in composition during storage. The substance should have a high equivalent weight so that the weighing errors may be negligible and it should be readily soluble under the conditions in which it is employed. The reaction with the primary standard solution must be stoichiometric and practically instantaneous. Usually an ideal primary standard is difficult to obtain and a compromise between the ideal requirements stated is generally necessary. The substances commonly used as primary standards are: sodium carbonate, oxalic acid, potassium dichromate, sodium tetraborate etc.

### 4.6.6 Normal Solution

A normal solution is defined as a solution containing one gram equivalent of the defined species or active reagent dissolved in one litre of solution. Equivalent weight expressed in grams is the gram equivalent. If the number of gram equivalent is 1/10, 1/100 or 1/1000 then the solution is designated as N/10 (decinormal), N/100 or 0.01 (N) *i.e.,* centinormal and N/1000 or 0.001 (N) (millinormal).

$$\text{Normality (N)} = \frac{\text{Number of grams of solute in 1000 mL of solution}}{\text{Equivalent weight of the solute}}$$

### 4.6.7 Molar Solution

A molar solution is one in which one gram molecular weight of the solute is dissolved in one litre of the solution. It is denoted by M. Thus a molar solution of oxalic acid ($C_2H_2O_4 \cdot 2H_2O$) contains $(2 \times 12 + 2 \times 1 + 4 \times 16 + 2 \times 18) = 126$ g per litre.

$$\text{Molarity (M)} = \frac{\text{Number of grams of solute in 100 mL of solution}}{\text{Molecular weight of the solute}}$$

### 4.6.8 Molal Solution

A molal solution is one which contains a gram molecular weight of the solute dissolved in 1000 grams of the solvent.

### 4.6.9 Formal Solution

A formal solution is one which contains a formula weight of a solute in a litre of the solution. It is denoted by F. In most of the cases formula weights and molecular weights are identical but sometimes the true molecular weight of a compound is a multiple of the weight expressed by its formula as ordinarily written in a chemical reaction.

$$\text{Formality (F)} = \frac{\text{Number of grams of solute per litre of solution}}{\text{Formula weight of solute}}$$

### 4.6.10 Factor of Solution

The factor of a solution is a number with which the strength of the proposed standard solution is to be multiplied to indicate the actual strength of the prepared solution. It is not always possible to weigh out exact amount of solute to prepare a solution of exact strength. In fact an amount nearest to the weight required is accurately weighed. For example to prepare 1 litre of (N/10) $Na_2CO_3$ solution 5.3 g is required. Let the actual weight taken be 5.45 g. Now,

5.3 g of $Na_2CO_3$ when present in 1 litre, the strength is (N/10)

5.45 g of $Na_2CO_3$ when present in 1 litre, the strength is 5.45/5.3(N/10) or 1.028(N/10). Hence, strength of the prepared solution is 1.028 (N/10),

Here 1.028 is the normality factor.

$$\text{Therefore factor of a solution (F)} = \frac{\text{Weight actually taken}}{\text{Weight required to be taken}}$$

### 4.6.11 Parts Per Million (ppm)

The concentration is expressed in terms of grams of solute per million millilitres of solution or milligrams of solute per litre of solution. Thus a solution containing *x* mg/litre of solute or *x* microgram (μg) of solute per millilitres of solution is *x* ppm solution.

### 4.6.12 Percentage Composition by Weight

The concentration is expressed in terms of grams of solute per 100 g of solution. For example a 10% KCl solution is prepared by dissolving 10 g of the salt in 90 g of water.

### 4.6.13 Percentage Composition by Volume

The concentration is expressed in terms of volumes of the solute and solvent.

### 4.6.14 Theory of Acid Base Titrations

Neutralisation indicators are substances which exhibit different colours according to the $H^+$ ion concentration or pH of the solution to which they are added. It is thus possible to have an idea of the pH of a given solution by adding a little of suitable indicator to the same. Moreover if an indicator is present during the progress of a titration of an alkali with an acid, the colour change of the indicator reveals the end point of titration. Most of the indicators have a predominantly 'acid colour' and a predominantly 'alkaline colour' in the lower and higher ranges of pH. The chief characteristics of such indicators is that the change from a predominantly 'acid' colour to a predominantly 'alkaline colour' is not sudden and abrupt, but takes place within a small interval of pH, termed as colour change interval of the indicators. For most acid-base titration we can therefore select an indicator which exhibits a distinct colour change at a pH close to that obtaining at the equivalence point.

Indicators are either very weak organic acids or bases which always exists in two tautomeric forms. One of the tautomers is in the non electrolytic forms and scarcely ionises but the other is an electrolyte and hence ionisable.

Let HIn′ be the non-ionisable tautomeric form and HIn be the ionisable form of phenolpthalein where $In^-$ represents the indicator ion. The colour of HIn and $In^-$ are the same but different from HIn′. In aqueous solution there would exist two equilibrium.

$$HIn' \rightleftharpoons HIn; \qquad K_t = \frac{a_{HIn}}{a_{HIn'}} \qquad ...(4.6.14.1)$$

$$HIn \rightleftharpoons H^+ + In^- ; \qquad K_D = \frac{a_{H^+} \cdot a_{In^-}}{a_{HIn}} \qquad ...(4.6.14.2)$$

In acid solution the dissociation will be supressed and the whole of the indicator ion shall remain in undissociated form *i.e.,* as HIn′ and HIn, which have different colours. The indicators will be applicable only if the equilibrium constant $K_t$ of eq. (4.6.14.1) is small so that the undissociated indicator mostly exists as HIn′.

Multiplying (4.6.14.1) and (4.6.14.2),

$$K_t \cdot K_D = \frac{a_{H^+} \cdot a_{In^-}}{a_{HIn}} = K_{in} \qquad ...(4.6.14.3)$$

where $K_{in}$ is called the indicator constant. It has been assumed that the concentration (hence activity) of HIn is very small. Therefore $a_{Hin^-}$ is the activity of practically the entire undissociated form. The colour of HIn′ is the colour which the indicator will show in acid medium. In presence of alkali, $H^+$ ions will be removed as $H_2O$, dissociation will be almost complete and indicator will be present mostly as $In^-$. Hence indicator ion shall have 'alkali' colour.

Rewriting (4.6.14.3),

$$a_{H+} = K_{in} \frac{a_{HIn'}}{a In^-} \qquad ...(4.6.14.4)$$

$$a_{H+} = K_{in} \cdot \frac{C_{HIn'}}{C_{In^-}} \frac{\gamma_{HIn'}}{\gamma_{In^-}} \qquad ...(4.6.14.5)$$

or

$$pH = pK_{in} + \log \frac{C_{In^-}}{C_{HIn'}} + \log \gamma_{In^-} \; ; \; [\gamma_{HIn'} = 1] \qquad ...(4.6.14.6)$$

Log $\gamma_{In^-}$ may be evaluated with extended Debye-Huckel law, but for most purposes, it is small and can be neglected, so that

$$pH = pK_{in} + \log \frac{C_{In^-}}{C_{HIn'}} \quad ...(4.6.14.7)$$

or

$$pH = pK_{in} + \log \frac{C_{\text{ionised form with alkaline colour}}}{C_{\text{non-ionised form with acid colour}}} \quad ...(4.6.14.8)$$

It is seen that at a given pH, indicator will exist in a definite ratio of concentrations of ionised and non-ionised form. Both the forms are present in any pH, but human eye can discern the colour distinctly when one predominates. It has been found that the acid colour, namely that of HIn′, is detected when $\frac{C_{HIn'}}{C_{In^-}} > 10$

*i.e.*, when $pH = pK_{in} - 1$

and the alkaline colour can be detected when $\frac{C_{In^-}}{C_{HIn'}} > 10$, *i.e.*, $pH = pK_{in + 1}$

As we titrate acid with base, the pH changes. The indicator will change from one colour to another within the pH range ($pK_{in+1}$) to ($pK_{in-1}$) within two units of pH. Thus phenolpthalein has a $pK_{in}$ value of 8.96 and hence its colour change takes place in the pH range 7.96 to 9.96. Methyl red has a $pK_{in}$ value of 5.1, so its colour would change in the pH range 4.1 to 6.1.

### 4.6.15 Principle of Acidimetry and Alkalimetry

Acidimetry is the method of determining the strength of an acid by titrating with a standard solution (*i.e.*, of known strength) of an alkali. The method with the help of which the strength of an alkali solution is found out by titrating with a standard solution of an acid is alkalimetry.

The law of normality can be stated as: "Equal volumes of solutions of two reacting substances (acids and bases) of same strength expressed in normality exactly neutralises each other" *i.e.*, V volume of X normal solution of any acid will exactly neutralise V volumes of X normal solution of any base. The strength of any solution decreases with dilution. In fact, the strength and volume of a solution are inversely proportional.

Therefore, Volume (V) of a solution $\propto \frac{1}{\text{strength(s) of the solution}}$

If $V_1$ volumes of an acid of exact strength $S_1$ neutralise $V_2$ volumes of a base of strength-$S_2$, then

$$\frac{V_1}{V_2} = \frac{S_2}{S_1} \quad ...(4.6.15.1)$$

or

$$V_1 S_1 = V_2 S_2 \quad ...(4.6.15.2)$$

### 4.6.16 Indicators

A substance which indicates the 'end point' or completion of reaction is known as indicator. It is an auxiliary reagent which helps in the visual determination of the completion of titration. Generally three types of indicator are used in volumetric analysis.

#### Internal indicator

The indicators are added into the solution where the reaction occurs, which gives a clear visual change in the solution being titrated e.g., methyl orange, methyl red, phenolpthalein etc.

The internal indicators are also divided into the following groups according to their use in different types of reactions.

- *Indicators used in acid-alkali neutralisation reaction*, viz. methyl red, phenolpthalein, methyl orange, bromothymol blue etc.

These are normally known as hydrogen ion or acid-base indicators. Actually such indicators are organic dyes (weak organic acids or bases) which changes colours within limits with variation in pH-value of the solution to which it is added. The important characteristics of these indicators is that the change from a predominantly 'acid' colour to a predominantly 'alkaline' colour is not sudden and abrupt, but takes place within a small interval of pH termed as the colour change interval of the indicator. It is advisable, to select an indicator which exhibits a distinct colour change at a pH close to that obtaining at the equivalence for acid-base titration.

- *Indicators used in precipitation reactions.* In the titration of sodium chloride with silver nitrate, a small quantity of $K_2CrO_4$ solution is added to serve as indicator. At the end point the chromate ion combines with $Ag^+$ ions to form sparingly soluble $Ag_2CrO_4$ (brick red colour). Both AgCl and $Ag_2CrO_4$ are insoluble, but they are precipitated only when their respective solubility products are reached (AgCl: $1.2 \times 10^{-10}$, $Ag_2CrO_4$: $1.7 \times 10^{-12}$). As long as $Cl^-$ are present in sufficient concentrations in titration mixture, only AgCl is precipitated and no $Ag_2CrO_4$ precipitates. When all $Cl^-$ ions are removed as AgCl only then a red precipitate of $Ag_2CrO_4$ appears and gives a reddish or brownish end point.

#### Redox indicators

Redox indicator is a substance which posses a different colour in the oxidised form and a different colour in the reduced form. Some organic dye stuff belong to this class. In order to get a sharp colour change at the end point the indicator chosen for a particular titration must have its standard potential ($E^o$) values in between the standard potential of the oxidation-reduction systems being titrated against each other. Examples are diphenylamine, methelyne blue, diphenylamineazo sulphonic acid etc. (see Article 4.7.2).

#### External indicator

These are not added to the reacting medium. For example, potasium ferricyanide $K_3[Fe(CN)_6]$, is used as an indicator in the titration of potassium dichromate with ferrous sulphate in an acid medium $K_3[Fe(CN)_6]$ reacts with $Fe^{2+}$ ion and forms a deep blue colour compound of ferro-ferricyanide

$$2K_3[Fe(CN)_6] + 3Fe^{++} \longrightarrow \underset{\text{deep blue}}{Fe_3[Fe(CN)_6]_2}$$

The indicator gives blue colour with $Fe^{2+}$ ions, hence when all the ferrous ions are oxidised in solution, then it will not give the blue colour.

### Self indicator

When one of the reactants itself acts as an indicator by visual colour change, is called a self indicator or auto-indicator. $KMnO_4$ acts as an auto-indicator whenever titrated with oxalic acid or ferrous sulphate in presence of dilute $H_2SO_4$.

Furthermore there are adsorption and complex forming indicators. In iodometric titration starch solution is used as an indicator which forms a complex with $I_2$, which has a very dark blue colour.

### 4.6.17 Choice of Indicators

When equivalent amounts of strong acid (say HCl) and a strong base (say KOH) are mixed, the resulting solution has a pH near about 7.0 and any indicator may be used in such neutralisation titrations. On the other hand, if to a weak acid solution an equivalent amount of strong base is added, the resulting salt in solution undergoes hydrolysis and the solution becomes alkaline having a pH above 7.0. Hence in a titration of weak acid and strong base, indicators whose colour change occurs in a higher range such as phenolpthalein or thymol blue should be used. For titrating strong acid with weak base (say HCl with $Na_2CO_3$), the resulting salt suffers hydrolysis and the solution becomes acidic even when equivalent amounts are added. The indicators like methyl red, methyl orange may be satisfactorily used whose colour change takes place in the acidic pH range.

## 4.7 OXIDATION AND REDUCTION REACTIONS: ELECTRONIC INTERPRETATION

Oxidation is the process which results in the loss of one or more electrons by atoms or ions. Reduction is the process which results in the gain of one or more electrons by atoms or ions. A oxidising agent is one that gains electrons and is reduced, like potassium permanganate ($KMnO_4$), potassium dichromate ($K_2Cr_2O_7$), Iodine ($I_2$) etc. A reducing agent is one that looses electrons and is oxidised like oxalic acid ($H_2C_2O_4$), ferrous sulphate ($FeSO_4$), stannous chloride ($SnCl_2$), sodium thiosulphate ($Na_2S_2O_3$).

In all oxidation-reduction processes (or redox processes), there will be a reactant undergoing oxidation and one undergoing reduction, since the two reactions are complementary to one another and occur simultaneously — one cannot take place without the other. The reagent suffering oxidation is termed as the reducing agent or reductant whereas the reagent undergoing reduction is termed as the oxidising agent or oxidant. The study of the electron changes in the oxidant and reductant forms the basis of the ion-electron method for balancing ionic equations. The equation is first divided into two balanced, partial equations representing the oxidation and reduction respectively.

Furthermore, it must be kept in mind that the reaction take place in aqueous solution so that in addition to the ions supplied by the oxidant and reductant hydrogen ions ($H^+$) and hydroxide ions ($OH^-$) are also present which is utilised in balancing the partial ionic equation. The unit change in oxidation or reduction is a charge of one electron, which is denoted by *e*. To understand the principles involved let us consider the reaction between ferric chloride ($FeCl_3$) and stannous chloride ($SnCl_2$) in aqueous medium, where actually ferric chloride is reduced by stannous chloride in aqueous solution. Ferric ions get reduced to ferrous ions by gaining electrons given up by stannous ions, which in the process become stannic ion upon oxidation. The partial ionic equation for reduction process is

$$Fe^{3+} \longrightarrow Fe^{2+} \qquad ...(4.7.1)$$

and for oxidation is,

$$Sn^{2+} \longrightarrow Sn^{4+} \qquad ...(4.7.2)$$

The equation (4.33) and (4.34) must be balanced not only with regard to the number and kind of atoms, but also electrically, that is the not electric charge on each side must be the same. Equation (4.33) can be balanced by adding one electron to the left hand side and (4.34) by adding two electrons to the right hand side.

Thus

$$Fe^{3+} + e \longrightarrow Fe^{2+} \qquad ...(4.7.3)$$

$$Sn^{2+} \longrightarrow Sn^{4+} + 2e \qquad ...(4.7.4)$$

These partial equations are then multiplied by the coefficients which result in the number of electrons utilised in one reaction being equal to those liberated in the other.

$$2Fe^{3+} + 2e \longrightarrow 2Fe^{2+} \qquad ...(4.7.5)$$

$$Sn^{2+} \longrightarrow Sn^{4+} + 2e \qquad ...(4.7.6)$$

Adding (4.37) and (4.38), we get:

$$2Fe^{3+} + Sn^{2+} + 2e \longrightarrow 2Fe^{2+} + Sn^{4+} + 2e \qquad ...(4.7.7)$$

Now, cancelling the electrons common to both sides, the simple ionic equation is obtained i.e.

$$2Fe^{3+} + Sn^{2+} \longrightarrow 2Fe^{2+} + Sn^{4+} \qquad ...(4.7.8)$$

Since, all strong electrolytes are completely dissociated, hence only the ions actually taking part or resulting from the reaction need appear in the equation. Substances which are only slightly ionised such as water or which are sparingly soluble and thus yield only as small concentration of ions e.g., silver chloride and barium sulphate are in general written as molecular formulae because they are present mainly in the undissociated state. Equation (4.7.8) is an example to prove that oxidation and reductions occurs simultaneously. Ion-electron balance therefore can be performed based on the step mentioned below:

- Ascertain the products of the reaction.
- Construct partial equation for oxidising agent.
- Construct partial equation for reducing agent in the same way.

- Multiply each partial equation by a factor so that when the two are added the electrons just gets compensated.
- Add the partial equations and cancel out the substances which appear on both sides of the equation.

### Illustration

- Reduction of potassium permanganate by iron (II) sulphate in presence of dilute sulphuric acid.

The first partial equation for reduction is

$$MnO_4^- \longrightarrow Mn^{2+}$$

To balance it atomically, $8H^+$ is required.

$$MnO_4^- + 8H^+ \longrightarrow Mn^{2+} + 4H_2O$$

To balance electrically 5*e* is required on the left hand side

$$MnO_4^- + 8H^+ + 5e^- \longrightarrow Mn^{2+} + 4H_2O$$

The second partial equation for oxidation is

$$Fe^{2+} \longrightarrow Fe^{3+}$$

To balance this electrically one electron must be added to the right hand side or subtracted from the left hand side.

$$Fe^{2+} \longrightarrow Fe^{3+} + e$$

Now the gain and loss of electrons must be equal. One permanganate ion utilises 5 electrons and one iron (II) ion liberates 1 electron; hence the two partial equations must apply in the ratio of 1 : 5

$$MnO_4^- + 8H^+ + 5e^- \longrightarrow Mn^{2+} + 4H_2O$$

$$5Fe^{2+} \longrightarrow 5Fe^{3+} + 5e^-$$

on addition we get

$$MnO_4^- + 8H^+ + 5Fe^{2+} = Mn^{2+} + 5Fe^{3+} + 4H_2O$$

- Partial ionic equation for potassium dichromate ($K_2Cr_2O_7$) in presence of sulphuric acid.

$$Cr_2O_7^{2-} \longrightarrow Cr^{3+}$$

$$Cr_2O_7^{2-} + 14H^+ \longrightarrow 2Cr^{3+} + 7H_2O$$

To balance electrically, 6*e* is to be added to the left hand side.

$$Cr_2O_7^{2-} + 14H^+ + 6e^- \longrightarrow 2Cr^{3+} + 7H_2O$$

### 4.7.1 Redox Potential

When a metal is immersed in a solution containing its own ions, a potential difference is established between the metal and its solution, which is called the electrode potential of the metal. The electrode potential E for the electrode reaction

$$M \rightleftharpoons M^{+n} + ne$$

is given by

$$E = E°_{M/M^{+n}} - \frac{RT}{nF} \ln a_{M^{+n}} \quad ...(4.7.1.1)$$

where $a_{M^{+n}}$ is the activity of metal ions in solution and $E°_{M/M^{+n}}$ is a constant called standard electrode potential.

When $a_{M^{+n}} = 1, E = E°_{M/M^{+n}}$ ...(4.7.1.2)

Thus standard electrode potential of a metal is the potential difference existing between the metal and a solution of its own ion of unit activity.

It is impossible to measure directly the electrode potentials. Only the electromotive force (e.m.f.) of a voltaic cell arising from a combination of two electrodes can be directly measured, which is given as the arithmetical sum or difference of the two electrode potential depending upon their signs. If one of the electrode potential be accurately measured, that of the other may be calculated. The reference electrode arbitrarily chosen for this purpose is the standard hydrogen electrode. Hydrogen gas at 1 atm. pressure and at a temperature of 25°C is slowly bubbled over a platinised platinum electrode which is immersed in a solution of hydrogen ions of unit activity. By convention potential of the half cell reaction

$$\frac{1}{2} H_2(p = 1 \text{ atm}) \rightleftharpoons H^+(a = 1) + e$$

is arbitrarily assigned the value of 0.00 volts. All other potential values (standard or otherwise) are referred to this value.

Thus standard electrode potential (oxidation) of zinc electrode is 0.763 volts. This means that for the voltaic cell,

$$Zn/Zn^{++}(a = 1)\ q/H^+\ (a = 1)\ /\ H_2\ (p = 1 \text{ atm})$$

cell e.m.f. is 0.763 volts.

The net cell reaction is

$$Zn + 2H^+\ (a = 1) \longrightarrow Zn^{++}\ (a = 1) + H_2\ (p = 1 \text{ atm})$$

The standard cell e.m.f. is

$$E° = E°_{Zn/Zn^{++}} - E°_{1/2H_2H^+} = E°_{Zn/Zn^{++}} = 0.763 \text{ V} \quad ..(4.7.1.3)$$

In a system containing both an oxidizing agent and its reduction product, there will be an equilibrium between them and the electrons.The inherent tendency of the redox system towards electron gain (*i.e.,* reduction) or less (*i.e.,* oxidation) can be measured as an electrical driving force and expressed as a potential value, called redox potential. The oxidation potential E of the redox process,

Red Ox + $ne$ is given by

$$E = E^\circ - \frac{RT}{nF} \ln \frac{a_{ox}}{a_{red}} \quad ...(4.7.1.4)$$

where E° is a constant called the standard oxidation potential (SOP). Evidently when $a_{ox} = a_{red}$ = 1, E = E°, a being the activities. Thus, SOP of a redox system is the potential difference between a platinum electrode and solution containing both the oxidized and the reduced form, each at unit activity, relative to standard hydrogen electrode at 25°C. Oxidation potential of a system measures the relative ease with which the reduced form of the redox couple is oxidised. Thus the oxidant in any couple will oxidize the reductant in any couple of higher positive potential. This fact is utilized in estimation of organic carbon content by rapid titration method of Walkley and Black (for details see Chapter 2).

The standard oxidation potential (SOP) values of $Fe^{2+}/Fe^{3+}$ and $Cr^{3+}/Cr_2O_7^{2-}$ couples are,

$$Fe^{2+} - e \longrightarrow Fe^{3+}; \qquad E^\circ_{Fe^{2+}/Fe^{3+}} = -0.77 \text{ volt}$$

$$2Cr^{+3} + 7H_2O - 6e \longrightarrow Cr_2O_7^{2-} + 14H^+; \qquad E^\circ_{Cr^{3+}/Cr_2O_7^{2-}} = -1.36 \text{ volt}$$

From the data it is evident that dichromate will oxidize ferrous ion to ferric state in acid medium.

$$Cr_2O_7^{2-} + 6Fe^{2+} + 14H^+ \longrightarrow 2Cr^{3+} + 6Fe^{+3} + 7H_2O$$

### 4.7.2 Redox Indicator

Redox indicator is a substance which posses a different colour in the oxidised form and a different colour in the reduced form. Some organic dye stuff belong to this class. In order to get a sharp colour change at the end point the indicator chosen for a particular titration must have its standard potential (E°) values in between the standard potential of the oxidation-reduction systems being titrated against each other. Examples are diphenylamine, methelyne blue, diphenylamineazo sulphonic acid etc.

In a redox titration of a reductant with an oxidant there occurs a continuous change of potential of the solution due to a gradual change of concentration of the species involved in the reaction. It can be easily shown that the potential changes abruptly in the neighbourhood of equivalence point and is dependant upon the standard potentials of the two oxidation – reduction systems involved but independent of the concentrations. In general for the reaction a

$$a_{OxI} + b_{RedII} \longrightarrow b_{OxII} + a_{RedI} \quad ...(4.7.2.1)$$

the oxidation potential at the equivalence point is given by

$$E^\circ = \left( \frac{bE^\circ_{I} + aE^\circ_{II}}{a+b} \right) \quad ...(4.7.2.2)$$

where $E^\circ_I$ and $E^\circ_{II}$ are the standard oxidation potentials of the systems $Red_I/Ox_I$ and $Red_{II}/Ox_{II}$.

An oxidation-reduction indicator is a substance which can mark the sudden change in the oxidation potential in the neighbourhood of the equivalence point in a redox titration. The ideal redox indicator will be one with an oxidation potential intermediate between that of the solution titrated and that of the titrand. A redox indicator is a compound which exhibits different colours in the reduced and oxidised forms.

$$In_{red} \longrightarrow In_{ox}$$

At a potential E, the ratio of concentrations of the oxidised and reduced forms is given by

$$E = E^{o}_{In} - \frac{RT}{nF} \ln \left[ \frac{In_{ox}}{In_{red}} \right] \qquad ...(4.7.2.3)$$

where $E^{o}_{In}$ is the standard oxidation potential (strictly the formal potential, discussed later). If the colour intensities of the two forms are comparable a practical estimate of the colour change interval corresponds to change of the ratio $[In_{ox}]/[In_{red}]$ from 10 to 1/10. The interval of potential is thus;

$$E = \frac{(E^{0}_{In} \pm 0.059)}{n} \text{ volt at } 25\ ^{o}C \qquad ...(4.7.2.4)$$

For a colour change at the end point, $E^{o}_{In}$ should differ by about at least 0.15 V from the standard (formal) potentials of the other systems involved in the reaction.

The earliest known redox – indicator is diphenylamine used for titration of Fe(II) with $K_2Cr_2O_7$. An intense blue violet colouration is produced at the end point. The action of diphenylamine(I) depends upon its oxidation first into colourless diphenylbenzidine(II) which is the real indicator and is reversibly further oxidised to diphenylbenzidine(III), (violet). Formal oxidation potential of (II)/(III) system is $E'^{o}_{In} = -0.76$ volt and $n = 2$

Hence the potential range of colour change of the indicator is $(-0.76 \pm 0.059/2) = -0.73$ volt to –0.79 volt. Above –0.73 volt the reduced form is predominant and solution is colourless while at –0.79 volt or below the oxidised form predominates and the solution assumes blue-violet colour. Now standard oxidation potential (E°) of $Fe^{2+}/Fe^{3+}$ and $2Cr^{3+}/Cr_2O_7^{2-}$ systems are –0.77 and –1.33, and the sharp fall of potential near the equivalence point occurs in the range –0.944 volt to –1.302 volt. Evidently diphenylamine is not a suitable indicator. However addition of phosphoric acid complexes Fe(III) as $[Fe(HPO_4)]^+$ ion. This increases the formal potential of Fe(II)/Fe(III) system so that the equivalence point potential coincides more nearly with that of the indicator. The limits of sharp change of potential in the titration curve (–0.712 volt to –1.302 volt), thus embraces the range of potential for colour change of diphenylamine which then functions quite suitable as an indicator. Use of barium diphenylamine sulphonate is to some extent preferable, firstly because it is water soluble and secondly its formal oxidation potential (–0.85 V in 0.5M $H_2SO_4$) is somewhat lower than that of diphenylamine.

### 4.7.3 Formal Potential

The standard oxidation potential E° of the redox system,

$$Red \longrightarrow Ox + ne$$

As given by Nernst equation,

$$E = E^\circ - \frac{RT}{nF} \ln \frac{a_{ox}}{a_{red}} \qquad ...(4.7.3.1)$$

is evaluated by taking complete considerations of the activities of the relevant species and with all the ions present in simple form. These are really limiting or ideal values which are rarely observed in actual experimental conditions. In practice the solutions may be quite concentrated when the activity of the pertinent species are much smaller than the concentrations, more so in the presence of other electrolytes. Besides, the actual active species present may differ from simple ions due to complexation. Evidently the use of standard potential data would not be very reliable. Under actual prevailing conditions it has been proposed that standard potentials be replaced by formal potentials. The formal potential is the potential observed experimentally in a solution containing equal number of moles of the oxidised and the reduced species together with other specified substances at specified concentrations. It takes into account the effects resulting from variation of activity coefficients with ionic strength, electrolytic dissociations, complexations, liquid junction potentials etc. The modified Nernst equation becomes

$$E = E^{\prime\circ} - 0.059 \log \left[\frac{a_{ox}}{a_{red}}\right] \text{ at } 25\ ^\circ\text{C} \qquad ...(4.7.3.2)$$

where $E^{\prime\circ}$ is the formal potential and $[a_{ox}]$ and $[a_{red}]$ are the molecular concentrations of the corresponding species.

Formal potentials vary appreciably with the nature and concentration of the present acid. Thus for $Fe^{2+}/Fe^{3+}$ system, the standard oxidation potential is –0.77 volt whereas formal potential $E^{\prime\circ}$ = –0.73 volts in 1(M)$HClO_4$, –0.68 volt in 1(M)$H_2SO_4$ and –0.61 volt in [0.5(M)$H_3PO_4$ + 1(M)$H_2SO_4$]. Evidently complexation is least in perchloric acid and greatest in phosphoric acid.

## 4.8 EQUIVALENT WEIGHT

The equivalent weight of a substance is defined as the weight of an element or compound which combines with or displaces from combination 8.00 parts by weight of oxygen, or 1.008 parts by weight of hydrogen or 35.45 parts by weight of chlorine.

### 4.8.1 Variability in Equivalent Weight

The equivalent weight of an element is decided on the basis of combination in its compounds. An element having variable valency can combine with an element with single valency to form more than one compound. As a result the equivalent weight of the element in the compounds will be different since the ratio of the weights of the two elements in the compounds are different. But if the element has only one valency then it would form only one compound with the second element *i.e.*, the ratio of the weight of the two elements in the compounds would be fixed. Hence the equivalent weight of the element will also be fixed. Many elements like Cu, Fe, N, P etc. show more than one valency. The equivalent weight of such elements, therefore, vary from one compound to the other. For example, iron forms two oxides viz. FeO (ferrous oxide) and $Fe_2O_3$ (ferric oxide). In FeO the equivalent weight of iron is $56 \times 8/16 = 28$ and in $Fe_2O_3$ it is $2 \times 56 \times 8/3 \times 16 = 18.33$.

### 4.8.2 Equivalent Weight and Valency

Let V and E be respectively the valency and equivalent weight of an element and A be the atomic weight.

Now one atom of the element will combine with V atoms of hydrogen, since valency is the number of hydrogen atoms which an atom of the element combine with.

The weight of V atoms of hydrogen is V (Since atomic weight of hydrogen = 1)

Now,1 part by weight of hydrogen combines with $\frac{A}{V}$ parts of the element

Therefore, $\frac{A}{V}$ is the equivalent weight of the element

*i.e.,*

$$E = \frac{A}{V} \text{ or } A = E \times V$$

*i.e.,* Atomic weight = equivalent weight × valency.

When V = 1, A = E *i.e.,* for monovalent elements atomic weight is same as the equivalent weight.

### 4.8.3 Equivalent Weight of Acid, Base and Salt

The equivalent weight of an acid is the number of parts by weight of the acid containing one part by weight of replaceable hydrogen. Hence,

$$\text{Equivalent wt. of an acid} = \frac{\text{Molecular weight of the acid}}{\text{No. of replaceable hydrogen atoms in one molecule of the acid}}$$

$$= \frac{\text{Molecular weight of the acid}}{\text{Basicity of the acid}}$$

The equivalent weight of a base may be defined as the number of parts by weight of the base that neutralise one equivalent of an acid.

Equivalent weight of a base

$$= \frac{\text{Molecular weight of the base}}{\text{The number of hydroxyl group present in one molecule of the base}}$$

$$= \frac{\text{Molecular weight of the base}}{\text{Acidity of the base}}$$

The equivalent weight of a salt may be defined as the number of parts by weight of the salt containing one equivalent part of the metal present in the salt. Thus equivalent weight of a salt may be obtained by dividing the molecular weight of the salt by the total valency of the metal atom or atoms present in one molecule of the salt.

Hence equivalent weight of a salt

$$= \frac{\text{Molecular weight of the salt}}{\text{The number of metal atoms in one molecule of the salt} \times \text{valency of the metal}}$$

### 4.8.4 Gram-equivalent Weight of Acid, Base and Salt

$$\text{Gram-equivalent weight of an acid} = \frac{\text{Gram molecular weight of the acid}}{\text{Basicity of the acid}}$$

$$\text{Gram-equivalent weight of a base} = \frac{\text{Gram molecular weight of the base}}{\text{Acidity of the base}}$$

$$\text{Gram-equivalent weight of a salt} = \frac{\text{Gram molecular weight of the salt}}{\text{Total positive valency in the formula of the salt}}$$

### 4.8.5 Equivalent Weight of an Oxidant or Reductant

The equivalent weight of an oxidant or reductant is the mole divided by the number of electrons which one mole of the substance gains or loses in the reaction; e.g.

$$MnO_4^- + 8H^+ + 5e^- \longrightarrow Mn^{2+} + 4H_2O \qquad Eq. = \frac{MnO_4^-}{5} = \frac{KMnO_4}{5}$$

$$Cr_2O_7^{2-} + 14H^+ + 6e^- \longrightarrow 2Cr^{3+} + 7H_2O \qquad Eq. = \frac{Cr_2O_7^{2-}}{6} = \frac{K_2Cr_2O_7}{6}$$

$$Fe^{2+} \longrightarrow Fe^{3+} + e^- \qquad Eq. = \frac{Fe^{2+}}{1} = \frac{FeSO_4}{1}$$

$$C_2O_4^{2-} \longrightarrow 2CO_2 + 2e^- \qquad Eq. = \frac{C_2O_4^{2-}}{2} = \frac{H_2C_2O_4}{2}$$

$$SO_3^{2-} + H_2O \longrightarrow SO_4^{2-} + 2H^+ + 2e^- \qquad Eq. = \frac{SO_3^{2-}}{3} = \frac{Na_2SO_3}{2}$$

$$2S_2O_3^{2-} \longrightarrow S_4O_6^{2-} + 2e^- \qquad Eq. = \frac{S_2O_3^{2-}}{1} = \frac{Na_2S_2O_3}{1}$$

### 4.8.6 Milliequivalents Per Litre

A solution containing milligram equivalent (1/1000 g equivalent) of a substance in a litre of solution is expressed as meq/litre.

## 4.9 ATOMIC WEIGHT

The atomic weight of an element may be defined as a number that gives the mass of one atom of that element compared with mass of one atom of oxygen taken as exactly 16 or mass of one atom of carbon taken as exactly 12.

$$\text{Atomic weight} = \frac{\text{Mass of one atom of the element}}{\text{Mass of one atom of oxygen}} \times 16$$

$$= \frac{\text{Mass of one atom of the element}}{\text{Mass of one atom of carbon}} \times 12$$

## 4.10 MOLECULAR WEIGHT

The molecular weight of a substance is a number that gives the weight of one molecule of that substance compared with the weight of one atom of oxygen taken as exactly 16 (or one atom of carbon takes as exactly 12).

$$\text{Molecular weight} = \frac{\text{Wt. of one molecule of the substance}}{\text{Wt. of one atom of oxygen}} \times 16$$

$$= \frac{\text{Wt. of one molecule of the substance}}{\text{Wt. of one atom of carbon}} \times 12$$

Molecular weight of a substance can be obtained by adding atomic weights of the various atoms of which a molecule is composed. Thus the molecular weight of chlorine from its formula $Cl_2$ is weight of $2Cl^-$ atoms = 2 × 35.5 = 71.

### 4.10.1 Gram Molecule or Gram Mole

The weight of a substance in grams which is equal to its molecular weight is known as gram-molecule or gram mole. The molecular weight of water is 18. Therefore one gram-mole of water weighs 18 grams.

### 4.10.2 Molar Volume

A gram mole of any gas at S.T.P *i.e.,* standard temperature and pressure (273 K and 760 mm Hg); occupies a volume of 22.4 litres. This is the molar volume. Thus 2 gm of hydrogen or 32 gm of oxygen will occupy a volume of 22.4 litres at S.T.P.

### 4.10.3 Mole-concept

Gram molecule of any substance is considered as mole. Both gram molecule and gram atom can be represented by mole. The term 'mole' of a substance may be defined as the weight in grams which contain $6.023 \times 10^{23}$ molecules. In the case of mono atomic molecule one mole represents the weight in grams of the element which contains $6.023 \times 10^{23}$ atoms of the element. In chemical analysis and calculation the weights of reacting substances are considered through the molecules and atoms of the substances taking part in the chemical reaction. The term 'mole' indicates at the same time the weight of the reacting substances as well as the number of molecules present in them. Here lies the significance of mole.

The term mole may be considered as follows:

- 1 mole = gram-molecule = molecular weight expressed in gram
- For gases, 1 mole = 22.4 litres at S.T.P.
- 1 mole = $6.023 \times 10^{23}$ molecules or atom = Avogadro's number.

Besides, mole is also used to represent one gram ion and one Faraday of electricity.

Thus, one mole of oxygen is the amount of oxygen in grams which contains Avogadro's number of oxygen atoms *i.e.,* $6.023 \times 10^{23}$ atoms. One mole oxygen is the atomic weight of oxygen expressed in grams (*i.e.,* 16 gram of oxygen).

One mole of nitrogen molecules means one gram molecule of nitrogen or 28 gram of nitrogen or $6.023 \times 10^{23}$ molecules of nitrogen.

One mole of ammonium ions means one gram ion of ammonium ions or 18 grams of ammonium ions or $6.023 \times 10^{23}$ ammonium ions.

## 4.11 MASS AND WEIGHT

The 'mass' of a substance is a definite property which can be used as a measure of quantity. The 'weight' of the substance is the force which results from the interaction of the gravitational force on the substance. Weight is thus the product of mass and acceleration due to gravity.

## 4.12 AVOGADRO'S HYPOTHESIS AND AVOGADRO'S NUMBER

Avogadro's hypothesis (1811) states "Equal volumes of all gases under S.T.P. contain equal number of molecules". Thus according to this hypothesis if one litre of hydrogen gas contains *n* molecules at 0°C and 760 mm of Hg pressure, one litre of any other gas (e.g., ammonia, chlorine, oxygen, nitrogen, carbon dioxide etc.) would contain n molecules under the same conditions of temperature and pressure. Experimental results have verified the above stated fact.

Avogadro's number is the number of molecules present in one gram-molecule of a substance. One gram molecule of any gas at S.T.P. occupies a volume of 22.4 litres. So Avogadro's number also denotes the number of molecules present in 22.4 litres of any gas at 0°C and 760 mm of Hg pressure. This number is a constant and its value is $6.023 \times 10^{23}$. This means that molecular weight of any substance expressed in grams contains $6.023 \times 10^{23}$ molecules or similarly atomic weight of any substance expressed in grams contain $6.023 \times 10^{23}$ atoms. Thus 1 gram-molecule hydrogen (2 g) and 1 gram molecule oxygen (32 g) though differ in weight, each of them contains equal number of molecules *i.e.,* $6.023 \times 10^{23}$.

●●●

# SUGGESTED READINGS


Alderfer, R.B. and Merkle, F.G. (1941). The Measurement of Structural Stability and Permeability and the Influence of Soil Treatment on the Properties. *Soil Sci.* **51**:201–212.

Amacher, M.C. (1984). Determination of Ionic Activities in Soil Solutions and Suspension: Principal Limitations, *Soil Sci. Soc. Am Proc* **48**:519–524.

American Public Health Association, Amer. Public Health Assoc. Amer. Water Works Assoc. and Water Pollution Control Fed. (APHA-AWWA-WPCF)(1985), Washington, D.C.

Atkins, P.W. (1986). *Physical Chemistry*, Third edition, Oxford University Press, Oxford 0 x 2 6DG.

Barua, T.C. and Barthakur, H.P. (1999). *A Textbook of Soil Analysis*, Vikas Publishing House Pvt. Ltd., New Delhi – 110 014.

Baver, L.D. (1956). *Soil Physics*, Third Ed., John Wiley & Sons, New York.

Black, C.A. (Ed.) (1965). *Methods of Soil Analysis*. Part I and II. American Society of Agronomy, Inc., Publishers, Madison, Wisconsin, USA.

Brady, N.C. (1984). *Nature and Properties of Soils*. MacMillan Co., New York.

Bray, R.H. and Kurtz, L.T. (1945). Determination of Total, Organic and Available Forms of Phosphorus in Soils. *Soil Sci.* **59**:39–45.

Chao, T.T., and Sanzolone, R.F. (1989). Fractionation of Soil Selenium by Sequential Partial Dissolution. *Soil Sci. Soc. Am.J.* **53**:385–392.

Chapman, J.F., and Dale, L.S. (1979). Atomic Absorption Spectrometric Determination of Some Elements Forming Volatile Hydrides with a Heated Cell Atomizer and Gas Handling System. *Anal. Chim. Acta* **111**:137–144.

Cooke, T.D., and Bruland, K.W. (1987). Aquatic Chemistry of Selenium: Evidence of Biomethylation. *Environ. Sci. Technol.* 221:1214–1219.

Chopra, S.L. and Kanwar, J.S. (1976). *Analytical Agricultural Chemistry*, Kalyani Publishers, Ludhiana.

Dickman, S.R. and Bray, R.H. (1940). Colorimetric Determinations of Phosphate. Indus & Engg. *Chem. Anal.* Ed. **12**:665–668.

Doran, J.W., and Alexander, M. (1977). Microbial Transformation of Selenium. *Appl. Environ. Microbial,* **3**:31–37..

Dyal, R.S. and Hendricks, S.B. (1950). Total Surface Area of Clays in Polar Liquids as a Characteristics Index, *Soil Sci*. **69**:421–432.

Ensminger, L.E. (1954). Some Factors Affecting the Adsorption of Sulphate by Alabama Soils. *Proc. Soil Sci. Soc. Am*. **18**:259–262.

Griffin, R.A. and Jurinak, J.J. (1973). Estimation of Activity Coefficients from Electrical Conductivity of the Natural Aquatic Systems and Soil Extracts. *Soil Sci*. **116**:26–30.

Gupta, U.C. (1967). A Simplified Method for Determining Hot Water Soluble Boron in Podzol Soils. *Soil Sci*. **103**:424–428.

Hesse, P.R. (1971). *A Textbook of Soil Chemical Analysis*. John Murray (Publishers) Ltd. London.

Holdgate, M.W. (1979) *A Perspective of Environmental Pollution*. Cambridge University Press, Cambridge.

Jackson, M.L. (1973). *Soil Chemical Analysis*, Prentice Hall Pvt. Ltd., New Delhi.

Jalota, S.K., Khera, R., and Ghuman, B.S. (1998). *Methods in Soil Physics*, Narosa Publishing House, New Delhi–110 017.

Jenkinson, D.S. (1988). The Determination of Microbial Carbon and Nitrogen in Soil, p. 368–386. In J.R. Wilson (ed.). Advances in Nitrogen Cycling in Agricultural Ecosystems. C.A.B. International, Wallingford.

John, M.K., Chuah, H.H. and Ndu feld. J.H. (1975). Application of Improved Azomethine-H Method for the Determination of Boron in Soils and Plants. *Anal. Lett*. **8**:559-568.

Johnston, E.S. and Barnard, W.M. (1979). Comparative Effectiveness of Fourteen Solutions for Extracting Arsenic from Western New York Soils. *Soil Sci. Soc. Am. J.* **43**:304–308.

Jones, U.S. (1982). *Soil Fertility and Fertilizers*. Reston Publ. Co., Reston, VA.

Kelley, W.P. (1948). *Cation Exchange in Soils*, Reinhold Publ. Corp., New York.

Lindsay, W.L. and Norvell, W.A. (1978). Development of a DTPA Test for Zn, Fe, Mn and Cu. *Soil Sci. Soc. Am. J.* **42**:421–428.

Masscheleyn, P.H., Delaune, R.D., and Patrick W.H., Jr. (1991a). Arsenic and Selenium Chemistry as Affected by Sediment Redox Potential and pH.J. *Environ. Qual*. 20:522–527.

McNeal, J.M., and Balistrieri, L.S. (1989). Geochemistry and Occurrence of Selenium: An Overview, p.1-13. In L.W. Jacobs *et al.* (ed.) Selenium in Agriculture and the Environment. SSSA Spec. Publ. 23. ASA and SSSA, Madison. WI.

McQuaker, N.R., Kluckner, P.D. and Chang, G.N. (1979). Calibration of an Inductively Coupled Plasma-atomic Emaission Spectrophotometer for the Analysis of Environmental Materials. *Anal. Chem*. **51(7)**: 888–895.

Miller, G.T. (1991) *Environmental Science: Sustaining the Earth*, Wadsworth Publishing, Belmont. C.A.

Olsen, S.R., Cole, C.V., Watnabe, F.S. and Dean, L.A. (1954). Estimation of Available Phosphorus in Soils by Estraction with Sodium Bicarbonate. U.S. Dep. Agric. Circ. 939.

Page, A.L. (1991). Methods of Soil Analysis, 2nd Edn. *Am.Soc. Agron. & Soil Sci. Am*. Madison, Wisconsin, USA.

Piper, C.S. (1950). *Soil and Plant Analysis*, Academic Press, New York.

Rattan, R.K., Datta, S.P., Chandra, Suresh and Saharan, Nilam (2002). Heavy Metals and Envrionmental Quality: Indian Scenario, *Fertilizers News*, **47**(11), pp.21–26 & 29–40.

Richards, L.A. and Weaver, L.R. (1943). Fifteen Atmosphere Percentage as Related to the Permanent Wilting Percentage. *Soil Sci.* **56**:331–340.

Richards, L.A. (1949). Pressure-membrance Apparatus, Construction and Use, *Agr. Eng.* **28**:451–454.

Richards, L.A. and Weaver, L.R. (1964). Moisture Retention by Some Irrigated Soils as Related to Soil Moisture Tension. *J. Agric. Res.* **69:**215–235

Russell, E.W. (1938). Measurement of Pore-space and Crumb or Aggregate Structure of Soils. Proc. Third Conf. Cotton Growers-problems, Rothamsted Exp. Stn.

Sanyal, S.K. and Dhillon, K.S. (2006) Arsenic and Selenium Dynamics in Water-Soil-Plant System: A Threat to Environmental Quality. *ICSWEQ Proceedings*, ISSS, New Delhi.

Sanzolone, R.F., and Chao T.T. (1987). Determination of Selenium in Thirty-two Geochemical Reference Materials by Continuous-flow Hydride Generation Atomic Absorption Spectrophotometry. *Geostand. Newslett.* **11**:81–85.

Schollenberger, C.J., and Dreibelbis F.R. (1930). Analytical Methods in Base-exchange Investigations in Soils. *Soil Sci.* **30**:160–173.

Schollenberger, C.J., and Simons R.H. (1945). Determination of Exchange Capacity and Exchangeable Bases in Soils. *Soil Sci.* **50**:13–24.

Sharma, S., and Singh R. (1983). Selenium in Soil, Plant and Animal Systems. p. 23–50. *In* C.P. Straub (ed.) *Crit. Rev. Environ. Control.* Vol. 13.

Shoemaker, H.E., McLean, E.O. and Pratt, P.F. (1961). Buffer Methods for Determining Lime Requirement of Soils with Appreciable Amounts of Extractable Aluminium. *Proc. Soil Sci. Soc. Am.* **25:**274–277.

Singh, R.A. (1980). *Soil Physical Analysis,* Second Edition. Kalyani Publishers, New Delhi, Ludhiana.

Soltanpour, P.N., Johnson, G.W., Workman, S.M., Jones, J.B. Jr. and Miller, R.O. (1998). Advances in ICP Emission and ICP Mass Spectrometry, *Adv. Agron.* **64**:27–113.

Soil Survey Staff (1075). Soil Taxonomy: A Basic System of Soil Classification for Making and Interpreting Soil Surveys. USDA-SCS Agric. Handb.436. U.S. Gov. Print. Office, Washington, DC.

Sorenson, S.P.L. (1909). *C.r. des traw. du. Lab. Carlsberg*, **8**(1).

Subbiah, B.V. and Asija, G.L. (1956). A Rapid Procedure for the Determination of Available Nitrogen in Soils. *Curr. Sci.* **25**:259–260.

Ure, A.M., and Berrow, M.L. (1982). The Elemental Composition of Soils. In H.J.M. Bowen (ed.) *Environmental Chemistry*, Vol. 2. Royal Soc. Chem., London.

Van Bavel, C.H.M. (1949). Mean Weight Diameter of Soil Aggregates as a Statistical Index of Aggregation. *Soil Sci. Soc. Am. Proc.* **14**:20–23.

Van Bavel, C.H.M. (1953). Report of the Committee on Physical Analysis 1951–1953, *Soil Sci. Soc. Am. Proc.* **17**:416–418.

Veihmeyer, F.J. and Hendrickson, A.H. (1931). The Moisture Equivalent as Measure of the Field Capacity of Soils. *Soil Sci.* **32**:181–194.

Veihmeyer, F.J. and Hendrickson, A.H. (1949). Methods of Measuring Field Capacity and Permanent Wilting Percentage of Soils. *Soil Sci.* **68**:75–95.

Vogel, A.L. (1962). *A Textbook of Quantitative Inorganic Analysis*, 3rd edn. Longmans, UK.

Voroney, R.P., and Paul, E.A. (1984). Determination of $K_c$ and $K_n$ *in situ* or Calibration of the Chloroform Fumigation Incubation Method. *Soil Biol. Biochem.* **16**:9–14.

Voroney, R.P., Winter, J.P., and Gregorich, E.G., (1991). Microbe/Plant Soil Interactions, pp. 77–99. In D.C. Coleman and B. Fry (ed.). Carbon Isotope Techniques Academic Press, New York.

Walkley, A. and Black, I.A. (1934). An Examination of the Degtjareff Method for Determining Soil Organic Matter, and a Proposed Modification of the Chromic Acid Titration Method. *Soil Sci.* **34**:29–38.

●●●

# APPENDICES: I to XXVII

## APPENDIX–I

### *Some Fundamental Constants*

| Constant | Value |
|---|---|
| Speed of light (c) | $2.9979 \times 10^{10}$ cm/sec |
| Electronic charge (e) | $4.8028 \times 10^{-10}$ esu |
| Avogadro's number ($\check{N}_o$) | $6.023 \times 10^{23}$ mole$^{-1}$ |
| Boltzmann constant (k) | $1.3806 \times 10^{-23}$ J/K |
| Gas constant (R) | 8.3143 J/°K (mole) |
| Planck's constant (h) | $6.6262 \times 10^{-27}$ ergs-sec |
| Gram-molar volume at NTP | 22414 cm$^3$ |

## APPENDIX–II

### *Physical Properties of Soils*

| Soil texture | Bulk density (Mg m$^{-3}$) | Total pore space (m$^3$m$^{-3}$) | Field capacity (%) | Infiltration rate (cm hr$^{-1}$) |
|---|---|---|---|---|
| Sandy | 1.65 | 38 | 6 | 5 |
| | (1.55–1.80) | (32–42) | (6–12) | (2.5–25) |
| Sandy loam | 1.50 | 43 | 14 | 2.5 |
| | (1.40–1.60) | (40–47) | (10–18) | (1.3–7.6) |
| Loam | 1.40 | 47 | 22 | 1.3 |
| | (1.35–1.60) | (47–51) | (18–26) | (0.8–2.0) |
| Clay loam | 1.35 | 49 | 27 | 0.8 |
| | (1.35–1.40) | (47–51) | (23–31) | (0.25–1.5) |

*Contd...*

| | | | | |
|---|---|---|---|---|
| Silty loam | 1.30 | 51 | 31 | 0.25 |
| | (1.30–1.40) | (49–53) | (27–35) | (0.03–0.5) |
| Clay | 1.25 | 53 | 35 | 0.5 |
| | (1.25–1.30) | (51–53) | (31–39) | (0.01–0.1) |

# APPENDIX–III

## *Mensuration of Surfaces and Volumes*

| | |
|---|---|
| Circumference of a circle | $\pi$ × diameter |
| Area of a circle | $\pi/4$ × diameter$^2$ |
| Area of square, parallelogram, rectangle | base × height |
| Area of trapezium | ½ sum of two parallel sides × height |
| Area of regular polygon | ½ radius of inscribed circle × length of one side × number of sides |
| Area of a triangle | ½ base × altitude |
| Area of parabola | 2/3 base × altitude |
| Area of ellipse | $\pi/4$ major axis × minor axis |
| Area of cycloid | 3 × area of generating circle |
| Surface area of a sphere | $\pi$ × (diameter)$^2$ |
| Volume of cube | side$^3$ |
| Volume of sphere | $\pi/6$ (diameter)$^3$ |
| Volume of paraboloid | ½ volume of circumscribing cylinder |
| Volume of prism | area of base × altitude |
| Volume of cylinder | ($\pi/4$) × (diameter)$^2$ × height |
| Volume of pyramid | ½ × area of base × height |
| Length of an arc | radius of circle × angle subtended at the center in radians |
| One radian | 57.29° |
| Volume of cone | 1/3 × $\pi/4$(diamerer of base)$^2$ × height |

# APPENDIX–IV

## *Some Conversion Factors*

**Force**

1 dyne = $2.248 \times 10^{-6}$ lb = $10^{-5}$ Newton

1 lb = $4.448 \times 10^{5}$ dyne = 4.448 Newton

**Pressure**

1 dyne/cm$^2$ = $9.869 \times 10^{-7}$ atm

1 atm = $1.013 \times 10^{5}$ dyne/cm$^2$

1 atm = 760 torr

1 torr = $1.333 \times 10^{2}$ dyne/cm$^2$ = $1.316 \times 10^{-3}$ atm

**Charge**

1 Coulomb = $1.036 \times 10^{-5}$ F

1 Faraday = 96520 Coulombs

**Energy**

| | *J* | *erg* | *eV* | *cal* |
|---|---|---|---|---|
| 1 joule | 1 | $10^{7}$ | $6.242 \times 10^{10}$ | 0.2389 |
| 1 erg | $10^{-7}$ | 1 | $6.242 \times 10^{11}$ | $2.389 \times 10^{-9}$ |
| 1 eV | $1.602 \times 10^{-10}$ | $1.602 \times 10^{-12}$ | 1 | $3.827 \times 10^{-20}$ |
| 1 cal | 4.186 | $4.186 \times 10^{7}$ | $2.613 \times 10^{10}$ | 1 |

1 Btu = $1.055 \times 10^{10}$ erg = 252.0 cal = 777.9 ft–lb.

1 cal = 3.087 ft–lb.

1 eV per molecule = 235 053 cal/g–mole

**Mass**

1 g = $6.024 \times 10^{23}$ amu

1 amu = $1.660 \times 10^{-24}$ g

1 lb = 453.59 g

1 g = $2.2046 \times 10^{-3}$ lb.

**Gas constant, R**

R (K, mole) = 8.3143 J

= $8.3143 \times 10^{7}$ ergs

= 1.987 calories

= 0.082 lit-atm. = $8.20 \times 10^{-2}$ dm$^2$ atm

= 82.05 c.c atm.

## *S.I. Units*

**S.I.Base**

| Physical quantity | Name of unit | Symbol |
|---|---|---|
| Length | metre | m |
| Mass | kilogram | kg |
| Time | second | s |
| Electric current | ampere | amp |
| Temperature | Kelvin | K |

**Derived units**

| Physical quantity | Name of S.I. Unit | Symbol | Definition |
|---|---|---|---|
| Force | newton | N | $kg.m.s^{-2}$ |
| Pressure | pascal | Pa | $N.m^{-2}$ |
| Energy | joule | J | $kg.m^2.s^{-2}$ |
| Power | watt | W | $J.s^{-1}$ |
| Electric charge | coulomb | C | amp.s |
| Potential difference | volt | V | $J.amp^{-1}s^{-1}$ |
| Electric resistance | ohm. | Ω | $v.amp^{-1}$ |
| Electric conductance | siemens | S | $amp.V^{-1}$ |
| Area | square metre | | $m^2$ |
| Volume | cubic metre | | $m^3$ |
| Density | kg per cubic metre | | $kg\ m^{-3}$ |
| Velocity | metre per second | | $ms^{-1}$ |
| Acceleration | metre per second squared | | $ms^{-2}$ |
| Viscosity | pascal-second | | Pas |
| Surface tension | newton per metre | | $Nm^{-1}$ |
| Electric field | volt per metre | | $Vm^{-1}$ |
| Heat capacity | joule per kelvin | | $JK^{-1}$ |

### *Prefixes for Fractions and Multiples of S.I. Units*

| Fraction | $10^{-1}$ | $10^{-2}$ | $10^{-3}$ | $10^{-6}$ | $10^{-9}$ | $10^{-12}$ |
|---|---|---|---|---|---|---|
| Prefix | deci | centi | milli | micro | nano | pico |
| Multiples | 10 | $10^2$ | $10^3$ | $10^6$ | $10^9$ | $10^{12}$ |
| Prefix | deca | hecto | kilo | mega | giga | tera |

### Some Non-SI Units and SI Units

| | | | |
|---|---|---|---|
| Length | angstrom | Å | $10^{-10}$m |
| Length | Micron | μm | $10^{-6}$m |
| Volume | litre | l | $10^{-3}m^3$ |
| Force | dyne | dyn | $10^{-3}$N |
| Pressure | bar | bar | $10^3$Pa |
| Pressure | atmosphere | atm | 101325 Pa |
| Energy | erg | erg | $10^{-7}$J |

# APPENDIX–V

### Hydraulic Conductivity Ratings

| Rating | K (cm $hr^{-1}$) |
|---|---|
| Very slow | < 0.125 |
| Slow | 0.125 – 0.50 |
| Moderately slow | 0.50 – 2.00 |
| Moderate | 2.00 – 6.25 |
| Moderately rapid | 6.25 – 12.50 |
| Rapid | 12.50 – 25.00 |
| Very rapid | > 25.00 |

# APPENDIX–VI

### Soil Acidity Class

| | |
|---|---|
| Extremely acid | < 4.5 |
| Very strongly acid | 4.5 – 5.0 |
| Strongly acid | 5.1 – 5.5 |
| Moderately acid | 5.6 – 6.0 |
| Slightly acid | 6.1 – 6.5 |
| Neutral | 6.6 – 7.3 |
| Mildly alkaline | 7.0 – 8.0 |
| Strongly alkaline | 8.1 – 9.0 |
| Very strongly alkaline | > 9.0 |

| pH | Rating | Deduction |
|---|---|---|
| < 6.5 | Acidic reaction | Requires liming |
| 6.5 – 7.5 | Normal | No treatment |
| 7.5 – 8.5 | Saline/Calcareous | Requires leaching of soluble salts |
| > 8.5 | Alkaline | Requires gypsum application |

# APPENDIX–VII

### *Mean Activity Coefficients in Water at 298 K*

| m/m⁻ | KCl | $CaCl_2$ |
|---|---|---|
| 0.001 | 0.966 | 0.888 |
| 0.01 | 0.902 | 0.732 |
| 0.1 | 0.770 | 0.524 |
| 1.0 | 0.607 | 0.725 |

# APPENDIX–VIII

### *Specific Conductances of KCl Solutions*

| Concentration (eqv./litre) | L (mhos) | | |
|---|---|---|---|
| | 0°C | 18°C | 25°C |
| 0.01 | 0.0007751 | 0.0012227 | 0.0014114 |
| 0.10 | 0.007154 | 0.011192 | 0.012886 |
| 1.00 | 0.06543 | 0.0982 | 0.11173 |

# APPENDIX–IX

### *Conductivity of 0.01N KCl Solution at Different Temperatures*

| Temperature (°C) | Conductivity mmhos $cm^{-1}$ |
|---|---|
| 15 | 1.147 |
| 16 | 1.173 |
| 17 | 1.199 |
| 18 | 1.225 |
| 19 | 1.251 |
| 20 | 1.278 |
| 21 | 1.305 |

*Contd...*

| | |
|---|---|
| 22 | 1.332 |
| 23 | 1.359 |
| 24 | 1.386 |
| 25 | 1.413 |
| 26 | 1.441 |
| 27 | 1.470 |
| 28 | 1.498 |
| 29 | 1.526 |
| 30 | 1.554 |
| 31 | 1.583 |
| 32 | 1.611 |
| 33 | 1.639 |
| 34 | 1.668 |
| 35 | 1.696 |

# APPENDIX–X

## *Temperature Factors for Correcting Conductivity Data on Soil Extracts to 25°C*

| Temperature (°C) | Correction factor |
|---|---|
| 10 | 1.411 |
| 11 | 1.375 |
| 12 | 1.341 |
| 13 | 1.309 |
| 14 | 1.277 |
| 15 | 1.247 |
| 16 | 1.218 |
| 17 | 1.189 |
| 18 | 1.163 |
| 19 | 1.130 |
| 20 | 1.112 |
| 21 | 1.087 |
| 22 | 1.064 |
| 23 | 1.043 |
| 24 | 1.020 |
| 25 | 1.000 |
| 26 | 0.979 |
| 27 | 0.960 |

***Contd...***

| | |
|---|---|
| 28 | 0.943 |
| 29 | 0.925 |
| 30 | 0.907 |
| 31 | 0.890 |
| 32 | 0.873 |
| 33 | 0.858 |
| 34 | 0.843 |
| 35 | 0.829 |

# APPENDIX–XI

### *Conductivity and Salinity*

| Conductivity (mmhos) | Approx Salt Concentration (%) | Salinity |
|---|---|---|
| 0 – 0.8 | < 0.05 | Normal |
| 0.8 – 1.6 | 0.05 – 0.15 | Saline |
| 1.6 – 3.2 | 0.15 – 0.25 | Highly saline |

# APPENDIX–XII

### *Common Commercial Concentrations of Acids and Ammonium Hydroxide*

| Reagent | Mass w/w (approximate weight %) | Normality (approximate) | Molarity (approximate) | Density | Ml. required to prepare 1 l of 1 (N) solution |
|---|---|---|---|---|---|
| Acetic acid, glacial | 99.7 | 17.4 | 17.4 | 1.05 | 57.5 |
| Hydrochloric acid | 37.0 | 12.1 | 12.1 | 1.19 | 82.6 |
| Hydrofluoric acid | 48.0 | 27.6 | 27.6 | 1.15 | 36.0 |
| Nitric acid | 90.0 | 21.1 | 21.1 | 1.48 | 47.4 |
| Nitric acid | 70.0 | 15.7 | 15.7 | 1.41 | 63.7 |
| Perchloric acid | 70.0 | 11.6 | 11.6 | 1.67 | 86.2 |
| Perchloric acid | 60.0 | 9.2 | 9.2 | 1.54 | 108.7 |
| Phosphoric acid | 85.0 | 44.0 | 14.7 | 1.69 | 22.7 |
| Sulfuric acid | 95.0 | 35.6 | 17.8 | 1.84 | 28.1 |
| Ammonium hydroxide | 57.6 | 14.8 | 14.8 | 0.90 | 67.6 |

# APPENDIX–XIII

## *Standard Solutions Commonly Used in Volumetric Analysis*

| Name | Formula | Equivalent weight | Normality | Preparation |
|---|---|---|---|---|
| Alcoholic KOH solution | KOH in Ethanol | 56 | 1N | 56 g KOH in 95% Ethanol to 1 lit. |
| Copper sulphate solution | $CuSO_4.5H_2O$ | 249.7 | 0.1N | 25.0 g in 1 lit.+ 2 ml conc. $H_2SO_4$ |
| EDTA solution | $Na_2C_{10}H_{12}N_2O_8$ | 372 | 0.01N | 2 g Sodium salt in 1 lit. solution |
| Ferrous sulphate solution | $FeSO_4.7H_2O$ | 278 | 0.5N | 139 g in 1 lit. solution + 10 ml $H_2SO_4$ |
| Ferrous ammonium sulphate | $FeSO_4(NH_4)_2SO_4.7H_2O$ | 392 | 0.1N | 39.2 g in 1 lit. solution + 5 ml conc. $H_2SO_4$ |
| Hydrochloric acid | HCl | 36.5 | 0.1N | 11 ml in 1 lit. solution |
| Iodine solution | $I_2$ | 126.9 | 0.1N | 12.7 g + 20 g KI in 10 ml water dilute to 1 lit. |
| Oxalic acid | $H_2C_2O_4.2H_2O$ | 63.02 | 0.1N | 6.302 g in 1 lit. water |
| Potassium dichromate | $K_2Cr_2O_7$ | 49 | 0.1N | Dissolve 4.9 g in 1 lit. $H_2O$ |
| Potassium permanganate | $KMnO_4$ | 31.6 | 0.1N | Dissolve 3.2 g in 100 ml hot water and dilute to 1 lit. |
| Silver nitrate | $AgNO_3$ | 169.9 | 0.1N | 17.0 g in 1 lit. water |
| Silver nitrate | $AgNO_3$ | 169.9 | 0.025N | 4.25 g in 1 lit. water |
| Sodium carbonate | $Na_2CO_3$ | 53 | 0.1N | 5.3 g in 1 lit. water |
| Sodium hydroxide | NaOH | 40 | 0.1N | 4.0 g in 1 lit. water |
| Sodium thiosulphate | $Na_2S_2O_3.5H_2O$ | 248.2 | 0.1N | 24.82 g in 1 lit. water |
| Sulphuric acid | $H_2SO_4$ | 49 | 0.1N | 3.6 ml conc. $H_2SO_4$ in 1 lit. water |

# APPENDIX–XIV

## *Equivalent Wt. of Acids and Bases*

| Name | Formula | Mol. wt. | Basicity or acidity | Eq. wt. |
|---|---|---|---|---|
| **Acids** | | | | |
| Acetic acid | $CH_3COOH$ | 60 | 1 | 60 |
| Hydrochloric acid | HCl | 36.5 | 1 | 36.5 |
| Nitric acid | $HNO_3$ | 63 | 1 | 63 |
| Oxalic acid | $H_2C_2O_42H_2O$ | 126 | 2 | 63 |
| Perchloric acid | $HClO_4$ | 84.5 | 1 | 84.5 |
| Phosphoric acid | $H_3PO_4$ | 98 | 3 | 32.6 |
| Sulphuric acid | $H_2SO_4$ | 98 | 2 | 49 |
| **Bases** | | | | |
| Calcium hydroxide | $Ca(OH)_2$ | 74 | 2 | 37 |
| Potassium hydroxide | KOH | 56 | 1 | 56 |
| Sodium bicarbonate | $NaHCO_3$ | 84 | 1 | 84 |
| Sodium carbonate | $Na_2CO_3$ | 106 | 2 | 53 |
| Sodium hydroxide | NaOH | 40 | 1 | 40 |

# APPENDIX–XV

## *Standard Equivalency Factors*

| Volumetric factors | | Factor |
|---|---|---|
| 1 ml $NH_2SO_4$ or HCl | 0.014 g N | |
| | | 0.017 g $NH_3$ |
| | | 3.050 g $CaCO_3$ |
| | 3.028 g CaO | |
| | 0.020 g Ca | |
| | 0.098 g $BaCO_3$ | |
| | 0.042 g $MgCO_3$ | |
| | 0.040 g NaOH | |
| | 0.056 g KOH | |
| | 0.084 g $NaHCO_3$ | |

| | |
|---|---|
| | $\equiv$ 0.030 g $CO_3^{--}$ |
| | $\equiv$ 0.061 g $HCO_3^-$ |
| | $\equiv$ 0.022 g $CO_2$ |
| 1 ml N NaOH or KOH | $\equiv$ 0.060 g $CH_3COOH$ |
| | $\equiv$ 0.003088 g $P_2O_5$ |
| | $\equiv$ 0.001349 g P |
| 1 ml N $KMnO_4$ | $\equiv$ 0.0316 g $KMnO_4$ |
| | $\equiv$ 0.063 g $H_2C_2O_4.2H_2O$ |
| | $\equiv$ 0.045 g $H_2C_2O_4$ |
| | $\equiv$ 0.020 g Ca |
| 1 ml N $KMnO_4$ or | $\equiv$ 0.392 g $FeSO_4(NH_4)_2SO_4.\ 6H_2O$ |
| $K_2Cr_2O_7$ | $\equiv$ 0.278 g $FeSO_4.7H_2O$ |
| | $\equiv$ 0.1519 g $FeSO_4$ |
| | $\equiv$ 0.05585 g Fe |
| | $\equiv$ 0.07984 g $Fe_2O_3$ |
| | $\equiv$ 0.07184 g FeO |
| 1 ml N $K_2Cr_2O_7$ | $\equiv$ 0.04903 g $K_2Cr_2O_7$ |
| | $\equiv$ 0.003 g Carbon |
| 1 ml N $Na_2S_2O_3$ | $\equiv$ 0.224 g $Na_2S_2O_3.5H_2O$ |
| | $\equiv$ 0.127 g $I^-$ |
| | $\equiv$ 0.0355 g $Cl^-$ |
| | $\equiv$ 0.0799 g $Br^-$ |
| | $\equiv$ 0.025 g $CuSO_4.5H_2O$ |
| | $\equiv$ 0.063 g Cu |
| 1 ml N $AgNO_3$ | $\equiv$ 0.0585 g NaCl |
| | $\equiv$ 0.0355 g Cl |
| | $\equiv$ 0.107 g Ag |
| | $\equiv$ 0.0745 g KCl |
| | $\equiv$ 0.0535 g $NH_4Cl$ |

# APPENDIX–XVI

## *Indicators Most Commonly Used*

| Name | Colour change | Types of indicator | Preparation |
|---|---|---|---|
| Ammonium purpurate (Murexide) | Pink to purple | Redox | 0.5 g in 100 g $K_2SO_4$ |
| Bromothymol blue | Yellow to blue | Acid-alkali | 1% in ethyl-alcohol |
| Diphenylamine | Violet to colourless | Redox | 1 g in 100 ml $H_2SO_4$ |
| Eriochrome black T | Purple to green | Complexometry | 1% in methanol |
| Methylene blue | Blue to colourless | Redox | 1% aqueous solution |
| Methyl orange | Pink to yellow | Acid-alkali | 1% aqueous solution |
| Methyl red | Pink to yellow | Acid-alkali | 1% in 50% ethanol |
| Phenolphthalein | Colourless to pink | Acid-alkali | 1 g in 100 ml 95% alcohol |
| Potassium ferricyanide | Colourless to brown | Precipitation | 4 g in 100 ml $H_2O$ |
| Starch | Colourless to blue | Adsorption | 1% in hot water. |
| Universal indicator | Several colours (at different pH) | pH indicator | 0.3 g Bromothymol blue<br>0.05 g Methyl orange<br>0.15 g Methyl red<br>0.35 g Phenolphthalein<br>Dissolve in 66% ethanol make 1 lit.with ethanol. |

# APPENDIX–XVII

## *pH Ranges of Various Indicators*

| Name of indicator | pH range | Colour in acid medium | Colour in alkaline (moderate) medium |
|---|---|---|---|
| Methyl orange | 3.1 – 4.5 | Red | Yellow |
| Methyl red | 4.2 – 6.3 | Red | Yellow |
| Bromothymol blue | 6.0 – 7.6 | Yellow | Blue |
| Phenol red | 6.4 – 8.2 | Yellow | Red |
| Litmus | 5.0 – 8.0 | Red | Blue |
| Phenolphthalien | 8.0 – 10.0 | Colourless | Pink |
| Thymolphthalein | 9.3 – 10.5 | Colourless | Blue |

# APPENDIX–XVIII

## *Selection of Indicator*

| Types of titration | Examples of reaction | Suitable indicator | pH range of colour change |
|---|---|---|---|
| Strong acid and strong base | NaOH v/s HCl, NaOH v/s $H_2SO_4$ KOH v/s HCl, etc. | Methyl orange, methyl red, bromothymol blue, Phenolphthalein | 5 to 10 |
| Strong acid and weak base | HCl v/s $Na_2CO_3$, HCl v/s $NaHCO_3$, HCl v/s $Ca(OH)_2$, $H_2SO_4$ v/s $Na_2CO_3$, $H_2SO_4$ v/s NaOH, etc. | Methyl orange, methyl red, (Not phenolphthalein) | 4 to 6 |
| Weak acid and strong base | $CH_3COOH$ v/s NaOH $CH_3COOH$ v/s KOH | Phenolphthalein, thymolphthalein (Not methyl orange or methyl red). | 8 to 10 |
| Weak acid and weak base | $CH_3COOH$ v/s $Na_2CO_3$ $CH_3COOH$ v/s $NaHCO_3$ | No suitable indicator. (Titration can not be done by using any indicator but potentiometric titration can be done) | Slow change of pH |

# APPENDIX–XIX

## *Some Redox Indicators*

| Indicator | Colour | | $E°$ at $H^+ = 1$ (volt) |
|---|---|---|---|
| | Oxidized form | Reduced form | |
| Neutral red | Red | Colourless | +0.24 |
| Indigo tetrasulfonate | Blue | Colourless | +0.36 |
| Methylene blue | Blue | Colourless | +0.53 |
| Diphenylamine | Blue–violet | Colourless | +0.76 |
| Diphenylamineazosulphonic acid | Red–violet | Colourless | +0.85 |
| Phenylanthranilic acid | Red–violet | Colourless | +1.08 |
| Ferroin ($Fe^{++}$ – orthophenanthroline) | Pale blue | Red | +1.14 |
| Nitroferroin | Pale–blue | Red | +1.25 |
| Diphenylamine–2, 2–dicarboxylic acid | Blue–violet | Colourless | +1.26 |

# APPENDIX–XX

## *Rating Chart for Soil Test Data*

| Fertility | Organic carbon (%) | Available N kg/ha | Available $P_2O_5$ kg/ha | Available $K_2O$ kg/ha |
|---|---|---|---|---|
| High | > 0.75 | > 450 | > 90 | > 340 |
| Medium | 0.50 – 0.75 | 280 – 450 | 45 – 90 | 150 – 340 |
| Low | < 0.50 | < 280 | < 45 | < 150 |

# APPENDIX–XXI

## *Critical Limits of Some Secondary and Micronutrients*

| Nutrients | Soil limits |
|---|---|
| S | <10 ppm hot water soluble |
| Ca | Below 50% of CEC ($NH_4OAc$ extractable) |
| Mg | Below 4% of CEC ($NH_4OAc$ extractable) |
| Zn | Below 0.6 ppm (DTPA extractable) |
| Fe | 2.5-4.5 ppm (DTPA extractable) |
| Mn | Below 2 ppm (DTPA extractable) |
| Cu | Below 0.2 ppm (DTPA extractable) |
| Cl | Below 2.0 ppm water soluble |

# APPENDIX–XXII

## *Standard Solution Preparation (100 ppm stock) for DTPA Extractable Fe, Mn, Zn and Cu*

| Element | Atomic weight | Salt | Mol. wt. | Weight of salt required in one litre |
|---|---|---|---|---|
| Zn | 65.38 | $ZnSO_4.7H_2O$ | 287.50 | 0.4398 |
| Cu | 63.54 | $CuSO_4.5H_2O$ | 249.69 | 0.3929 |
| Mn | 54.94 | $MnCl_2.4H_2O$ | 197.91 | 0.3602 |
| Fe | 55.85 | $FeSO_4.7H_2O$ | 278.02 | 0.4977 |

# APPENDIX–XXIII

## *Standard Atomic Absorption Conditions of Some Elements in Air-Acetylene*

| Element | Wavelength ( λ ) nm | Slit | Sensitivity check* |
|---|---|---|---|
| Fe | 248.3 | 0.2 | 5.0 |
| Mn | 279.5 | 0.2 | 2.5 |
| Cu | 324.8 | 0.7 | 4.0 |
| Zn | 213.9 | 0.7 | 1.0 |
| Ca | 422.7 | 0.7 | 4.0 |
| Mg | 285.2 | 0.7 | 0.3 |

***Metal concentration (mg/l) in aqueous solution which will give a reading of approximately 0.2 absorbance units.**

# APPENDIX–XXIV

## *Rating for Carbonate and Bicarbonate*

| Soluble salts(%) in 0-20 cm soil layer | | Remarks |
|---|---|---|
| $CO_3^{2-}$ | $HCO_3^-$ | |
| Nil | <0.06 | Non-salinized |
| <0.005 | - | Weakly salinized |
| 0.05 – 0.10 | - | Average salinized |
| 0.011 – 0.030 | - | Strongly salinized |

# APPENDIX–XXV

## *Rating for Chloride*

| $Cl^-$ (%) in 0-20 cm layer of soils | Remarks |
|---|---|
| < 0.020 | Non-salinized |
| 0.20–0.50 | Weakly salinized |
| 0.51–0.120 | Average salinized |
| 0.121–0.200 | Strongly salinized |

# APPENDIX–XXVI

## *Stability Constants (log K) of Metal-EDTA Complexes*

| Ion | log K | Ion | log K |
|---|---|---|---|
| $Mg^{2+}$ | 8.7 | $Fe^{3+}$ | 25.1 |
| $Ca^{2+}$ | 10.7 | $Cr^{3+}$ | 24.0 |
| $Sr^{2+}$ | 8.6 | $Ce^{3+}$ | 15.9 |
| $Ba^{2+}$ | 7.8 | $La^{3+}$ | 15.7 |
| $Mn^{2+}$ | 13.8 | $Th^{4+}$ | 23.2 |
| $Fe^{2+}$ | 14.3 | $Ag^{+}$ | 7.3 |
| $Co^{2+}$ | 16.3 | $Li^{+}$ | 2.8 |
| $Nl^{2+}$ | 18.6 | $Na^{+}$ | 1.7 |
| $Cu^{2+}$ | 18.8 | | |
| $Zn^{2+}$ | 16.7 | | |
| $Cd^{2+}$ | 16.7 | | |
| $Hg^{2+}$ | 21.9 | | |
| $Pb^{2+}$ | 18.0 | | |
| $Al^{3+}$ | 16.3 | | |

# APPENDIX–XXVII

## *Some Important Conversion Factors*

Acre × 2.471 = hectare = $2.24 \times 10^6$ kg Soil (Plough layer)

Hectare × 0.404686 = Acre

Meq $L^{-1}$ × equivalent weight = ppm

ppm = $\mu g\ mL^{-1}$ = $mg\ L^{-1}$

ppm = $mg\ kg^{-1}$

N × 1.286 = $NH_4$

N × 4.43 = $NO_3$

Organic carbon × 1.724 = Organic matter

P × 2.29 = $P_2O_5$

K × 1.20 = $K_2O$

Ca × 1.40 = CaO

Ca × 1.85 = $Ca(OH)_2$

S × 3 = $SO_4$